Ivan Gutman · Sven J. Cyvin

Introduction to the Theory of Benzenoid Hydrocarbons

With 15 Figures

Springer-Verlag
Berlin Heidelberg New York
London Paris Tokyo Hong Kong

Professor Dr. **Ivan Gutman**
University of Kragujevac
Faculty of Science
P.O. Box 60
YU-34000 Kragujevac

Professor Dr. **Sven Josef Cyvin**

The University of Trondheim
The Norwegian Institute of Technology
Division of Physical Chemistry
N-7034 Trondheim-NTH

ISBN 3-540-51139-3 Springer-Verlag Berlin Heidelberg New York
ISBN 0-387-51139-3 Springer-Verlag New York Berlin Heidelberg

Library of Congress Cataloging-in-Publication Data
Gutman, Ivan, 1947 –
Introduction to the theory of benzenoid hydrocarbons/Ivan
Gutman, Sven Josef Cyvin.
  p. cm.
Includes bibliographical references.
ISBN 0-387-51139-3 (U.S.)
1. Polycyclic aromatic hydrocarbons. I. Cyvin, Sven J. (Sven Josef), 1931–   . II. Title.
QD341.H9G89   1989
547'.611—dc20   89-21714   CIP

© Springer-Verlag Berlin Heidelberg 1989
Printed in Germany

Printing: Saladruck, Berlin. Bookbinding: Lüderitz & Bauer, Berlin
2151/3020-543210 – Printed on acid-free paper

Preface

In the last hundred years benzenoid hydrocarbons have constantly attracted the attention of both experimental and theoretical chemists. In spite of the fact that some of the basic concepts of the theory of benzenoid hydrocarbons have their origins in the 19th and early 20th century, research in this area is still in vigorous expansion.

The present book provides an outline of the most important current theoretical approaches to benzenoids. Emphasis is laid on the recent developments of these theories, which can certainly be characterized as a significant advance. Emphasis is also laid on practical applications rather than on "pure" theory.

The book assumes only some elementary knowledge of organic and physical chemistry and requires no special mathematical training. Therefore we hope that undergraduate students of chemistry will be able to follow the text without any difficulty. Since organic and physical chemists are nowadays not properly acquainted with the modern theory of benzenoid molecules, we hope that they will find this book both useful and informative. Our book is also aimed at theoretical chemists, especially those concerned with the "topological" features of organic molecules.

The authors are indebted to Dr. WERNER SCHMIDT (Ahrensburg, FRG) for valuable discussions. One of the authors (I. G.) thanks the Royal Norwegian Council for Scientific and Industrial Research for financial support during 1988, which enabled him to stay at the University of Trondheim and write the present book.

Trondheim, July 1989   Ivan Gutman   Sven J. Cyvin

Contents

Chapter 1

Benzenoid Hydrocarbons

1.1 Acquaintance

Benzenoid hydrocarbons are condensed polycyclic unsaturated fully conjugated hydrocarbons composed exclusively of six-membered rings. Some of their representatives are depicted in Fig. 1.1. The structural formulas of benzenoid hydrocarbons are given in the usual abbreviated form:

The distribution of the π-electrons in a benzenoid hydrocarbon is not correctly represented by any (KEKULÉ-type) classical structural formula with double bonds between certain pairs of carbon atoms. This problem will be thoroughly examined in subsequent parts of the present book. In this chapter we simply write the structural formulas without any indication of the π-electrons.

In the current chemical literature the so-called *polycyclic aromatic hydrocarbons* (PAHs) are frequently mentioned. Therefore the distinction between the class of polycyclic aromatic hydrocarbons and the class of benzenoid hydrocarbons should be made clear at this early point.

Polycyclic aromatic hydrocarbons may contain rings with sizes different from six (usually five-membered rings) and/or carbon atoms which do not participate in the conjugated π-electron network (i.e. sp^3 hybridized carbon atoms) and/or side-groups.

Some examples of this type are fluoranthrene (1), fluorene (2), hexahydrobisanthene (3), and methylcoronene (4), presented in Fig. 1.2.

Benzenoid hydrocarbons are not allowed to possess the structural features mentioned above. On the other hand, a PAH is by definition aromatic; this stipulates a pronounced degree of thermodynamic and chemical stability. Some

naphthalene, $C_{10}H_8$
colorless, mp 80°C

anthracene, $C_{14}H_{10}$
colorless, mp 216°C

phenanthrene, $C_{14}H_{10}$
colorless, mp 101°C

pyrene, $C_{16}H_{10}$
colorless, mp 150°C

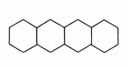

naphthacene, $C_{18}H_{12}$
orange-red, mp 356°C

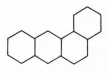

tetraphene, $C_{18}H_{12}$
colorless, mp 161°C,
carcinogenic

chrysene, $C_{18}H_{12}$
colorless, mp 254°C,
carcinogenic

benzo[c]phenanthrene, $C_{18}H_{12}$
colorless, mp 66°C,
carcinogenic

triphenylene, $C_{18}H_{12}$
colorless, mp 198°C

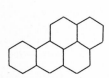

benzo[a]pyrene, $C_{20}H_{12}$
pale yellow, mp 178°C,
carcinogenic

benzo[e]pyrene, $C_{20}H_{12}$
colorless, mp 179°C

perylene, $C_{20}H_{12}$
orange-yellow, mp 273°C

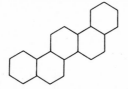

triangulene, $C_{22}H_{12}$

unstable, never obtained

picene, $C_{22}H_{14}$

colorless, mp 367°C,
carcinogenic

coronene, $C_{24}H_{12}$

pale yellow, mp 438°C

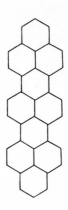

heptacene, $C_{30}H_{18}$

green-black, very unstable

terrylene, $C_{30}H_{16}$

red-violet, mp 510°C

teropyrene, $C_{36}H_{18}$

red-purple, mp > 500°C

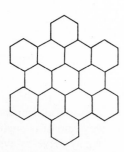

hexabenzo[bc,ef,hi,kl,no,qr]coronene,

$C_{42}H_{18}$, orange-yellow, mp > 700°C

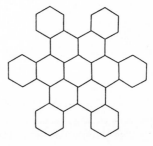

hexabenzo[a,d,g,j,m,p]coronene,

$C_{48}H_{24}$, yellow, mp 516°C

Fig. 1.1. Examples of benzenoid hydrocarbons. About 300 such hydrocarbons have been synthesized and experimentally characterized until now, which is only a minute fraction of their theoretically possible number (see Chapter 4)

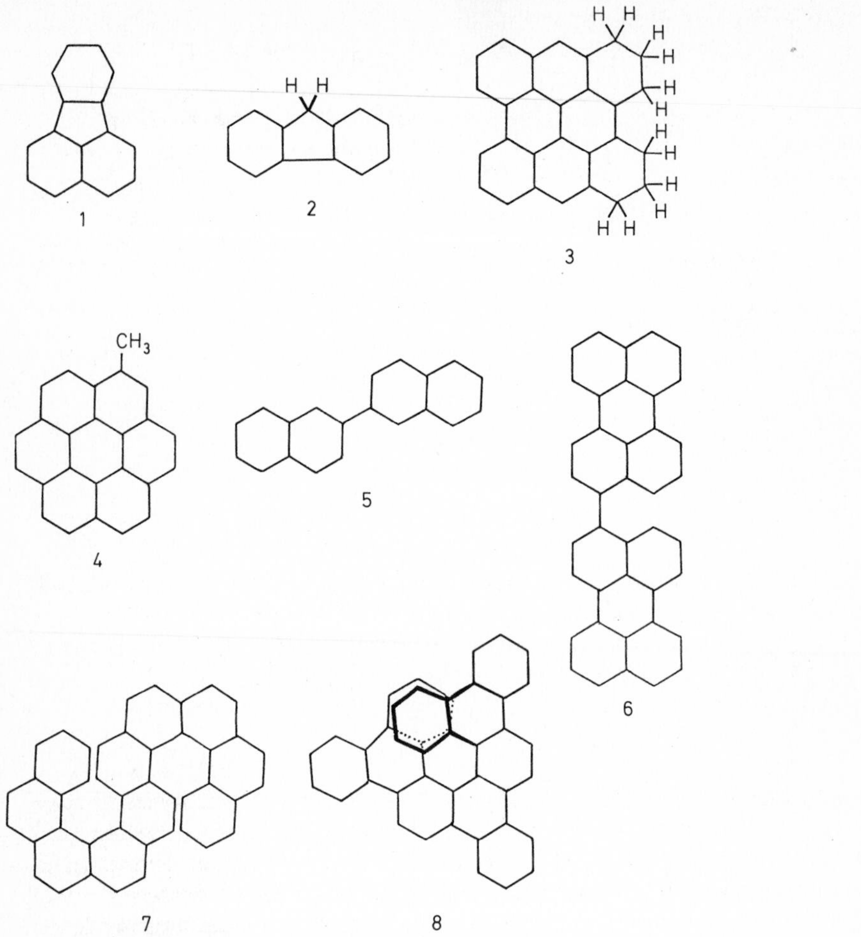

Fig. 1.2. Polycyclic aromatic hydrocarbons which are not benzenoid hydrocarbons

benzenoid hydrocarbons are highly reactive compounds, or even transient species which perhaps exist as reaction intermediates, and thus they cannot be included among PAHs. Heptacene and triangulene (see Fig. 1.1) may serve as typical examples.

These latter examples are, however, exceptions. The vast majority of the benzenoid hydrocarbons studied experimentally so far are reasonably stable compounds and thus belong to the class of PAHs.

Polycyclic aromatic hydrocarbons like 2,2′-dinaphthyl (5) or diperylenyl (6) are not described as benzenoids because they are not condensed in the strict sense of the word.

As a final restriction, the molecules of benzenoid hydrocarbons must be planar or, to be less exclusive, nearly planar. Thus helicenic compounds such as diphen-

anthro[3,4-*c*:3′,4′-*l*]-chrysene (7) or tribenzo[*f,l,r*]heptahelicene (8) are not considered to be benzenoid hydrocarbons, at least not in this book.

It is not at all obvious why 7 and 8 are not benzenoid hydrocarbons, whereas
benzo[*c*]phenanthrene (Fig. 1.1), whose nonplanarity is experimentally established
(HERBSTEIN and SCHMIDT 1954), and dibenzo[*c,g*]phenanthrene, whose enantio-
meric forms have been separated (MIKEŠ et al. 1976), are accepted as benzenoids.

enantiomers of dibenzo[c,g]phenanthrene

The motivation behind the exclusion of helicenic systems from this book will
become clear in the next chapter where the mathematical objects called "benzenoid
systems" are introduced. Then the requirement that a benzenoid molecule must
be "nearly planar" will get a more precise meaning.

*

After so many groups of polycyclic hydrocarbons have been declared beyond
the scope of the present book, it is worth asking how many benzenoid hydrocarbons
are actually known. It is not easy to answer this seemingly simple question because
the separation, isolation, purification, and identification of benzenoid hydrocarbons is very difficult. The problem is especially tough with compounds with
molecular masses above 300 because of the large number of isomers and their very
similar solubilities and volatilities (for details see LEMPKA et al. 1985).

The recent compilation by CIOSLOWSKI and WALA (1986) mentions 307 benzenoid hydrocarbons for which at least some experimental data are available in
published form. In the handbook by DIAS (1987), experimental data are collected
for only 244 benzenoid and 18 helicenic hydrocarbons. The collection of WERNER
SCHMIDT (Ahrensburg, FR Germany) contains 242 benzenoid hydrocarbons that
are all well characterized stable substances.

Taken together, these sources indicate that the number of known benzenoid
hydrocarbons is around 300.

1.2 Nomenclature

Throughout this book the benzenoid hydrocarbons as well as their derivatives are
named according to the rules defined by IUPAC. These somewhat complicated

rules can be found elsewhere (e.g. in Chapter 1 of DIAS 1987). However, the reader of the present book will be able to follow the text easily even if he is not familiar with the IUPAC nomenclature.

1.3 Occurrence

Polycyclic aromatic hydrocarbons (and therefore also benzenoid hydrocarbons) are formed during pyrolysis or incomplete combustion of almost all kinds of organic materials. The product obtained is a dark-colored, viscous substance called tar or pitch*; its composition depends somewhat on the starting material, but much more on the conditions under which the pyrolysis has been performed, primarily the temperature (SCHMIDT 1987). Tars and pitches are extremely complicated mixtures (LAND and EIGEN 1967). It is believed that the coal tar contains

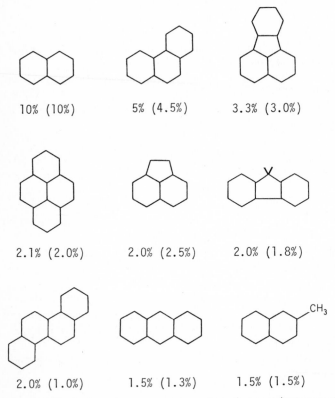

10% (10%)   5% (4.5%)   3.3% (3.0%)

2.1% (2.0%)   2.0% (2.5%)   2.0% (1.8%)

2.0% (1.0%)   1.5% (1.3%)   1.5% (1.5%)

Fig. 1.3. The hydrocarbons present in coal tar that constitute more than 1% of the whole (ULLMANN 1982); data in brackets are taken from FRANCK and STADELHOFER (1987)

---

* A tar is a liquid with relatively low viscosity at ordinary temperature. A pitch is a very viscous or almost solid substance.

some 10000 distinct compounds (FRANCK and STADELHOFER 1987). The main components of a typical coal tar are presented in Fig. 1.3.

As seen from Fig. 1.3, only five benzenoid hydrocarbons are present in tar in relatively large quantities. Small and trace amounts of many other benzenoids, however, are present not only in tar but practically everywhere in the environment. This is a serious pollution problem which is briefly discussed in Section 1.5.

Small quantities of benzenoid hydrocarbons can be found also in food, especially in products processed by frying or roasting.

Benzenoid hydrocarbons are the constituents of a number of minerals. The association of PAHs with mercury ores had already been noticed in the last century. A material called "Stupp" or "Stuppfat" would condense in the retorts during the "roasting" of the mercury ore from the mines of Idrija, Yugoslavia. In 1887 pyrene was identified as one of the components of "Stupp". Other minerals of a similar kind are idrialite (found in Idrija, Yugoslavia and elsewhere) and curtisite (found in California). They are composed of picene, chrysene, and smaller amounts of at least 30 PAHs (GEISSMAN et al. 1967, BLUMER 1975). More surprising is that the mineral pendletonite (found in California) is almost pure – 99% – coronene (MURDOCH and GEISSMAN 1967, BLUMER 1975).

Various asbestos minerals have been reported to contain benzenoid hydrocarbons and in particular the carcinogenic benzo[a]pyrene (SCHMIDT 1987).

Soils and young marine sediments contain various benzenoid hydrocarbons, among which are phenanthrene, anthracene, chrysene, pyrene, perylene, benzo[e]-pyrene, benzo[a]pyrene, coronene (for their formulas see Fig. 1.1), benzo[ghi]pery-lene, and anthanthrene (BLUMER 1976, SCHMIDT 1987):

benzo[ghi]perylene   anthanthrene

Benzenoid hydrocarbons have been discovered in quite unusual materials. Minor amounts of naphthalene, phenanthrene, anthracene, and pyrene were found in certain meteorites (PERING and PONNAMPERUMA 1971, HAHN et al. 1988). For instance, the phenanthrene content of the Murchison meteorite is 5.0 ppm (HAHN et al. 1988). Phenanthrene, perylene and anthracene, as well as the compound 3 from Fig. 1.2, were found in a 150 million year old fossil sea lily (BLUMER 1976). Quite recently the phenalene radical was detected in flint* (CHANDRA et al. 1988).

---

* Flint is an extremely hard form of polycrystalline silica.

phenalene, $C_{13}H_9$

Out of the numerous technical products which contain benzenoid hydrocarbons we mention here only carbon black. This material is used in the production of tyres and as printing ink. It is obtained by burning various hydrocarbons and contains some 99.6% carbon. Pyrene, benzo[a]pyrene, anthanthrene, benzo[ghi]-perylene, and coronene were found among the PAHs contained in carbon black.

More data on the occurrence of benzenoid hydrocarbons can be found in the review by SCHMIDT (1987). A few years ago, benzenoid hydrocarbons were detected in interstellar clouds. This topic is briefly outlined in Chapter 9.

1.4 Applications

If one disregards the direct uses of tars and pitches for various industrial and technical purposes, only two benzenoid hydrocarbons, namely naphthalene and anthracene, have large-scale applications in the chemical industry (ULLMANN 1982, FRANCK and STADELHOFER 1987).

Naphthalene is obtained from coal tar. About 65% of it is converted into phthalic anhydride (I) by catalytic gas-phase oxidation.

I   II   III

Phthalic anhydride is further used in the production of plastics. About 15% of naphthalene is converted into naphthalenesulfonic acid, β-naphthol, etc. and used for the production of dyes. Some naphthalene is used for moth balls.

Anthracene is obtained from coal tar. It is almost completely oxidized into anthraquinone (II) and used for synthetic dyes.

Very pure (99.9999%) samples of anthracene have been prepared by zone refining. They are used in nuclear physics as scintillation counters. Anthracene has also been suggested for use as a photoconductor and semiconductor.

Phenanthrene is the second major constituent of coal tar. Part of it is oxidized into diphenic acid (III) and further used for polymers. However, the production of phenanthrene exceeds the demands of the chemical industry, and therefore significant amounts of it have simply to be left unisolated in tar. An efficient method for transforming phenanthrene into its more valuable isomer anthracene is not known in spite of recent efforts (COLLIN and ZANDER 1983).

Pyrene is used in dye production (FRANCK and STADELHOFER 1987). Chrysene has found applications as a photosensitizer and for UV-filters (ULLMANN 1982). Coronene is used to convert UV to visible light in UV detectors (BLOUKE et al. 1980).

Technical and chemical uses of other benzenoid hydrocarbons are rather limited.

The nine compounds below are mentioned in the handbook of DIAS (1987) as being used as semiconductors (IV), constituents of oxygen gas detectors (V), photochromic additives for plastics (VI, VII, VIII), and photoconductors (IX–XII).

1.5 Benzenoid Hydrocarbons as Pollutants

The fact that workers exposed to soot and tar (e.g. chimney sweeps) more frequently than other people develop certain sorts of cancer was observed in the 18th century (SEARLE 1986). In 1933 benzo[a]pyrene was identified as the carcinogenic constituent of coal tar (COOK et al. 1933). A number of other benzenoid hydrocarbons also are reported to be carcinogenic (for examples, see Fig. 1.1).

Having this in mind, the ubiquitous occurrence of benzenoid hydrocarbons in the products obtained by heating and burning organic materials is a serious worldwide acute health problem (SEARLE 1986).

A certain amount of PAHs is produced in some natural processes (e.g. volcanic activity, forest fires, formation of soil), but anthropogenic sources of PAHs are by far the most important ones as far as the pollution of the human environment is concerned.

In the United States alone, the annual emission of polycyclic aromatic hydrocarbons is estimated to be 6000 tons. Automobiles contribute to this by approximately 36%, industrial production by 28%, residential heating by 12%, power generation by 7%, etc.

It is estimated that automobiles produce between 5 and 50 µg of benzo[a]pyrene per kilogram of gasoline. If a catalyst is used, then the benzo[a]pyrene production may fall to only 0.4 µg/kg. A jet airplane releases 2–10 mg benzo[a]pyrene per minute into the environment.

Among various industries, by far the greatest PAH polluters are the aluminium producers. Some 15 g of benzo[a]pyrene or a total of 235 g of PAH may be released for each ton of aluminium!

Other major PAH polluters are iron works (60 g PAH/ton iron), coke oven plants (15 g PAH/ton coal charged), ferroalloy industry (10 g PAH/ton alloy), production of carbon black (0.3 g PAH/ton) etc.

The PAH emission of residential heating devices (stoves, furnaces etc.) lies around 1 g of benzo[a]pyrene for a ton of wood or coal. In the case of coal-fired power plants the PAH emission depends very much on the technology employed, the size of the unit and the way in which the plant is operated. Typical emission values are between 1 and 10 mg benzo[a]pyrene/ton coal.

All data in this section are taken from BJØRSETH and RAMDAHL (1985).

*

Tobacco smoke contains over 3800 compounds including benzo[a]pyrene and many other known carcinogens which are not PAHs (SEARLE 1986). Inhalation of tobacco smoke brings these substances directly into the lungs. According to a recent estimate, tobacco use accounts for over 30% of all cancer deaths in the USA (SEARLE and WATERHOUSE 1988).

For the mechanism by which benzo[a]pyrene interacts with DNA (which may be responsible for its carcinogenity) the reader should consult the review by HARVEY and GEACINTOV (1988) and the references cited therein.

Chapter 2

Benzenoid Systems

Benzenoid systems are geometric figures. They are composed of congruent regular hexagons, arranged according to certain rules.

Let us first get acquainted with the hexagon, the fundamental building block of the benzenoid systems.

A *hexagon* is a plane figure of six sides and six angles. If all the sides are equal and if all the angles are equal, then the hexagon is said to be *regular*. In the theory we are going to present, it is customary to call the side of a hexagon an *edge* and the point where two edges meet a *vertex*:

regular hexagon

In this book regular hexagons will always be drawn so that two of their edges are vertical.

Two regular hexagons are congruent if their edges are equal in size. All regular hexagons which occur in this book are assumed to be congruent.

There are several ways in which a benzenoid system can be defined. Before presenting these precise but boring definitions, it is instructive to look at a few examples. In Fig. 2.1 the benzenoid systems composed of 2, 3, and 4 hexagons are depicted. By inspecting Fig. 2.1 the reader may get a fairly clear idea about the nature of benzenoid systems.

Definition A. A benzenoid system is a connected geometric figure obtained by arranging congruent regular hexagons in a plane, so that two hexagons are either disjoint or have a common edge. This figure divides the plane into one infinite (external) region and a number of finite (internal) regions. All internal regions must be regular hexagons.

Definition B. Consider a regular hexagon χ and label its edges by $e_1, \ldots, e_6$, so that for $i = 1, \ldots, 5$, e_i and e_{i+1} have a common vertex.

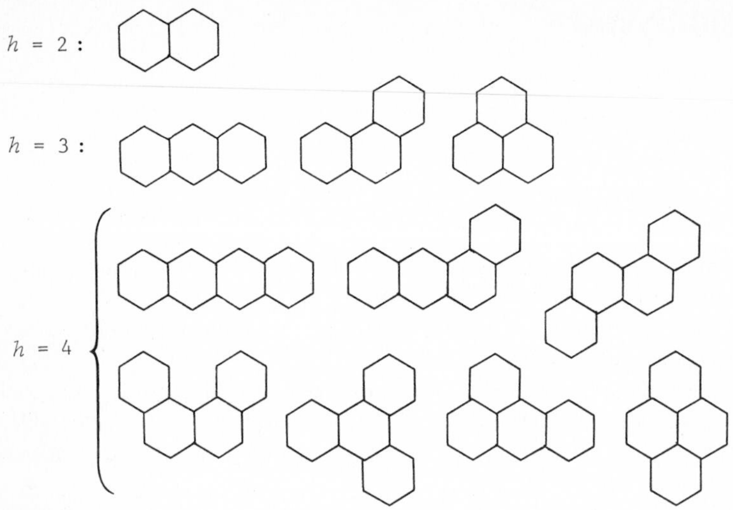

$h = 2:$

$h = 3:$

$h = 4$ {

Fig. 2.1. The only 11 possible, non-isomorphic, benzenoid systems with h hexagons, $h = 2, 3, 4$

Denote by h the number of hexagons in a benzenoid system. The only benzenoid system with $h = 1$ is a regular hexagon itself. For $h \geq 1$, a benzenoid system with $h + 1$ hexagons is a plane figure obtained by adding the hexagon χ to a benzenoid system B_h with h hexagons in one of the following five ways:

(i)   The edge e_1 of χ is identified with an edge of B_h.
(ii) The edges e_1 and e_2 of χ are identified with two edges of B_h.
(iii) The edges e_1, e_2, and e_3 of χ are identified with three edges of B_h.
(iv) The edges e_1, e_2, e_3, and e_4 of χ are identified with four edges of B_h.
(v)   The edges e_1, e_2, e_3, e_4, and e_5 of χ are identified with five edges of B_h.

All benzenoid systems with $h + 1$ hexagons are constructed by the above procedure.

The above five addition modes can be visualized as follows:

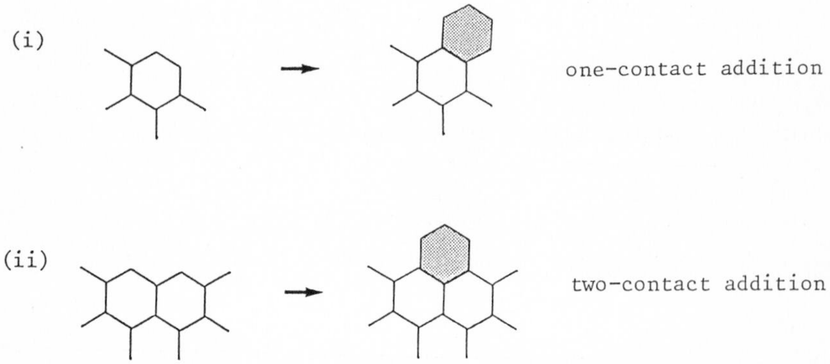

(i)   →   one-contact addition

(ii)   →   two-contact addition

(iii)

three-contact addition

(iv)

four-contact addition

(v)

five-contact addition

Only one of the six possible orientations of each of the above diagrams is depicted. The hexagon χ is hatched. The pendent lines symbolize hexagons which may, but need not, exist.

It is known that the plane can be completely covered by congruent regular hexagons. The infinite network formed in this way is called the hexagonal lattice. Chemists often use the name graphite lattice for the same object.

A cycle on the hexagonal lattice is a set of distinct edges $e_1, e_2, ... , e_p$ belonging to this lattice, such that for $i = 1, 2, ... , p - 1$, e_i and e_{i+1} have a common vertex and, in addition, e_1 and e_p have a common vertex. For an example see Fig. 2.2.

Definition C. Let C be a cycle on the hexagonal lattice. A benzenoid system is formed by the vertices and edges lying on C and in the interior of C.

A surprisingly large number of names have been given for what we just have defined as benzenoid systems. Some of the synonyms for "benzenoid system" are:

hexagonal animal	(HARARY 1967)
hexanimal	(HARARY and HARBORTH 1976)
polyhex	(BALABAN and HARARY 1968)
fusene	(BONCHEV and BALABAN 1981)
hexagonal polyomino	(GOLOMB 1965)
honeycomb system	(SACHS 1984)
hexagonal system	(SACHS 1984)
benzenoid graph	(TRINAJSTIĆ 1983, GUTMAN and POLANSKY 1986)

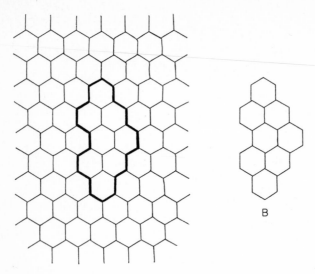

Fig. 2.2. The hexagonal lattice and a cycle on it. *B* is the benzenoid system determined by this cycle (cf. Definition C). The cycle is called the perimeter of *B*

As will be explained in a while, the meaning of "benzenoid graph" is somewhat different.

From a mathematician's point of view, the name "hexagonal system" or even "hexagonal polyomino" may be more appropriate. Bearing in mind the purpose of this book and wishing not to stray far from chemistry, we have chosen the name which directly relates to the intimate connection between our objects and benzenoid hydrocarbons.

Many authors consider benzenoid systems as graphs* (see e.g. Vol I, p. 23 of TRINAJSTIĆ 1983 and p. 59 of GUTMAN and POLANSKY 1986) and call them "benzenoid graphs". Indeed, the vertices and the edges of a benzenoid system form a graph. For example:

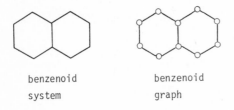

benzenoid   benzenoid
system   graph

---

* In a nutshell: a graph is a mathematical object consisting of some unspecified elements called "vertices" and a binary relation, defining which pairs of vertices are "adjacent" (and which are not). If a graph is represented by a diagram, then the vertices are drawn as points and adjacent vertices are connected by lines called "edges". For more detail on graphs see GUTMAN and POLANSKY (1986).

At first glance it is not easy to see whether there is any difference between a benzenoid system and the corresponding benzenoid graph. The differences are easier to grasp after observing that a graph, being an abstract mathematical structure, can have arbitrarily many pictorial (geometric) representations. For instance, the diagrams below all correspond to the same graph, that is, to the naphthalene graph from the previous example.

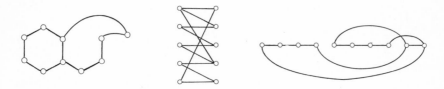

Obviously, a lot of geometric information is lost in the graphical representation of a benzenoid system. As a consequence, many notions which occur in the theory of benzenoid hydrocarbons (perimeter, peak, valley, monotonic path, segmentation etc.) cannot be properly defined in graph-theoretical terms.

This, however, does not mean that the graph-theoretical approach is of little use in the theory of benzenoid hydrocarbons. Just the contrary, whenever it is justified and profitable, one should exploit the power and mathematical elegance of chemical graph theory. Nevertheless, in this book we adhere to the geometric nature of the benzenoid systems and use graph-theoretical reasoning only where appropriate.

The three definitions of benzenoid systems given above are equivalent, but reveal different features of these objects. These definitions must, however, be supplemented by the following

Definition D. Denote by σ the plane to which the benzenoid systems belong. Two benzenoid systems are *isomorphic* (i.e. they are two copies of one and the same object) if they can be brought into coincidence by any combination of translations in σ, rotations in σ and a reflection in a plane perpendicular to σ.

This seemingly complicated definition guarantees that exactly one benzenoid system is associated with each benzenoid hydrocarbon. An example is given in Fig. 2.3.

Concerning the definition of benzenoid systems we wish to make a final remark. According to any of the definitions A–C, benzenoid systems are planar and consist of regular hexagons of equal size, which should not possess overlapping edges. As a consequence, there is no benzenoid system which would correspond to a (non-planar) helicenic molecule, e.g. to heptahelicene. Further, benzenoid systems should not contain "holes" and therefore cannot represent coronoid hydrocarbons, e.g. kekulene. Making a slight modification in Definitions A, B,

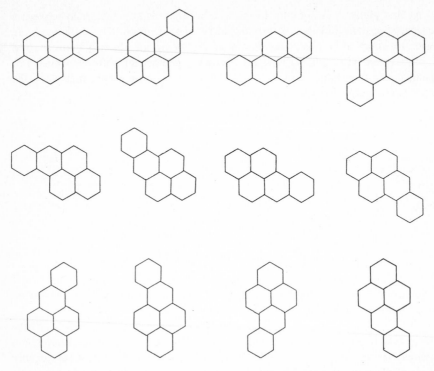

Fig. 2.3. Illustration of Definition D: Isomorphic benzenoid systems; they all represent benzo[a]pyrene

or C one can introduce the coronoid systems, i.e. benzenoid-like systems with holes. Their theory is outlined in detail in Chapter 8.

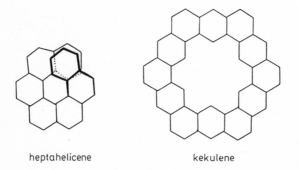

heptahelicene   kekulene

Interlude

There is an obvious correspondence between a benzenoid hydrocarbon and a benzenoid system. One example may suffice:

dibenzo[fg,op]naphthacene

a benzenoid system,
which we also call
"dibenzo[fg,op]naphthacene"

This close relation between the structure of a chemical compound (benzenoid hydrocarbon) and a mathematical object (benzenoid system) is, of course, the basis for the entire theory outlined in this book. We shall first exploit this relation by giving to a benzenoid system the same name as the corresponding benzenoid hydrocarbon.

Many other chemical notions have their complete analogues in the theory of benzenoid systems. Here we point out the coincidence between a KEKULÉ structure (of a benzenoid hydrocarbon) and a so-called 1-factor (of a benzenoid system):

Kekulé structures of a benzenoid hydrocarbon

1-factors of a benzenoid system

For a formal definition of a 1-factor see the next chapter.

Chapter 3

Anatomy

In this chapter, the most important structural characteristics of benzenoid systems are described and the basic relations between them pointed out. After learning enough about the anatomy of the benzenoid systems, we shall be able to put forward several classification schemes.

3.1 Symmetry

In this section we examine the symmetries of benzenoid systems. A regular hexagon belongs to the symmetry group D_{6h}. Benzenoid systems, being composed of regular hexagons, belong therefore either to the symmetry group D_{6h} or to a subgroup thereof that includes reflection in the horizontal plane. These subgroups are C_{6h}, D_{3h}, C_{3h}, D_{2h}, C_{2h}, C_{2v}, and C_s. One example for each symmetry is depicted below.

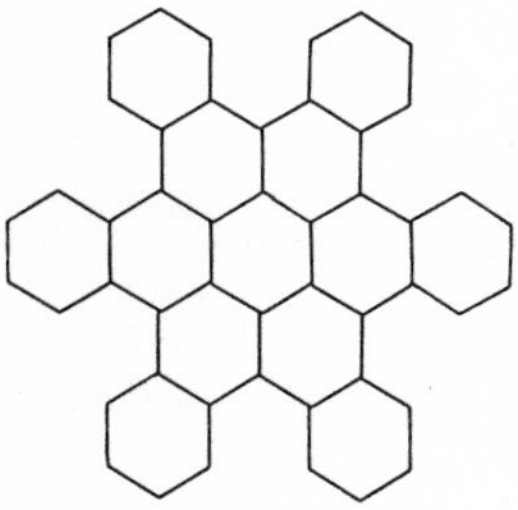

D_{6h}

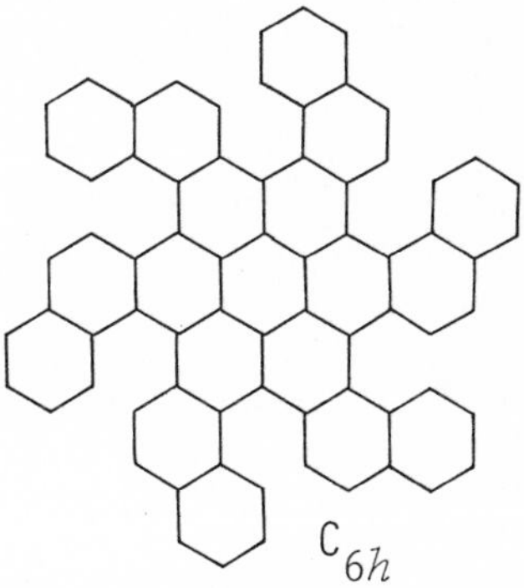

C_{6h}

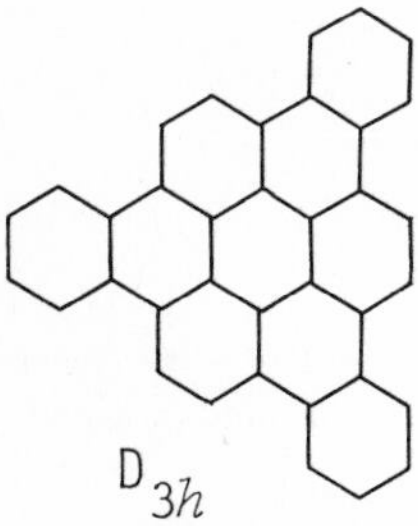

D_{3h}

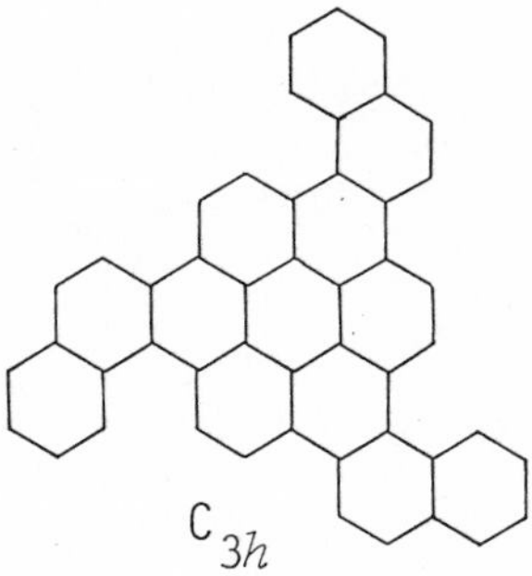

C_{3h}

D_{2h}

C_{2h}

C_{2v}

C_s

3.2 Vertices, Edges, and Related Concepts

As we already know, a benzenoid system contains *vertices* and *edges*. If two vertices are the endpoints of the same edge, then they are said to be *adjacent*. If two edges share a common vertex, then they are *incident*. Also, an edge and the vertex which is its endpoint are incident.

A synonym for "adjacent" is "neighboring". Two adjacent vertices are said to be *first neighbors* or simply *neighbors*.

We use the term adjacent also for hexagons: two hexagons are adjacent if they share a common edge.

Let v be a vertex. The number of first neighbors of v is the *degree* of the vertex v. It is obvious that the vertices of a benzenoid system have either degree two or degree three.

The notions defined above are illustrated on an example given in Fig. 3.1.

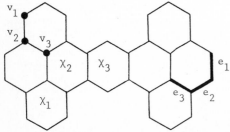

Fig. 3.1. A benzenoid system having $h = 9$ hexagons, $n = 34$ vertices, and $m = 42$ edges. The vertices v_1 and v_2 are adjacent; v_1 and v_3 are not adjacent. The edges e_1 and e_2 are incident; e_1 and e_3 are not incident. The hexagons χ_1 and χ_2 are adjacent; χ_1 and χ_3 are not adjacent. The vertices v_1, v_2, and v_3 are of degree 2, 3, and 3, respectively. The vertices v_1 and v_2 are external; v_3 is an internal vertex. The edges e_1 and e_2 are external; e_3 is an internal edge

With respect to the number of neighbors of a hexagon and their mutual positions, one can distinguish twelve distinct modes. These *modes of hexagons* are denoted by L_1, L_2, L_3, L_4, L_5, L_6, A_2, A_3, A_4, P_2, P_3, and P_4. Their definition is clear from Fig. 3.2, where an example is also provided.

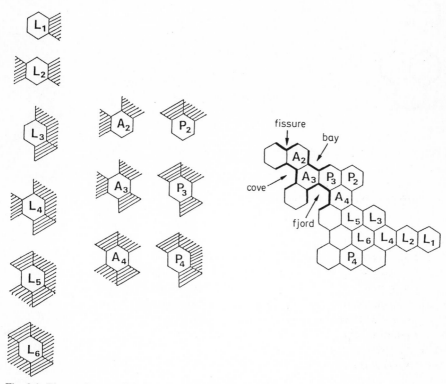

Fig. 3.2. The twelve possible modes of hexagons in benzenoid systems and the notation proposed for them. Definition of some structural features of the perimeter

Two adjacent hexagons share a common edge or, in other words, in a benzenoid system some edges are shared by two hexagons. These edges are called *internal*. Those edges which belong to just one hexagon are called *external*. The external edges form a cycle which is called the *perimeter* (or by some authors the boundary) of the benzenoid system. The perimeter concept is easily understood from Definition C given in the preceding chapter; from this definition it is clear that the cycle of the hexagonal lattice which determines the benzenoid system is just its perimeter; for illustration, see Fig. 2.2.

The *external vertices* of a benzenoid system are those lying on its perimeter. The other vertices are *internal*; it may happen that a benzenoid system has no internal vertices.

The *size of the perimeter* is equal to the number of external edges (or external vertices). Note that the perimeter is not necessarily the greatest cycle in a benzenoid system, as is illustrated in the example below:

A cycle of size 24 in a benzenoid system whose perimeter has the size 22

An external edge whose vertices are both of degree two is called a *free edge*. Every benzenoid system possesses at least six free edges.

At this point we introduce the names *fissure*, *bay*, *cove*, and *fjord* for certain details of the perimeter. Instead of a formal definition, the reader is referred to Fig. 3.2, from which these notions should be clear.

There exist numerous relations between the above defined structural character-istics of benzenoid systems. For example, it should be fairly obvious that all internal vertices have degree three, or that there is an equal number of external edges and external vertices. We now provide a complete list of equalities connecting the numbers of the pertinent structural details.

Let

$$
\begin{array}{ll}
h & = \text{number of hexagons} \\
n & = \text{number of vertices} \\
m & = \text{number of edges} \\
n_i & = \text{number of internal vertices}
\end{array}
$$

These quantities are related via

$$
\begin{aligned}
n &= 4h + 2 - n_i , \\
m &= 5h + 1 - n_i , \\
m &= n + h - 1 .
\end{aligned}
$$

The third equation is obtained, of course, by subtracting the first from the second. In what follows, # will have the meaning "number of". Then we have:

$$
\begin{aligned}
\# \text{ external vertices} &= 4h + 2 - 2n_i , \\
\# \text{ vertices of degree two} &= 2h + 4 - n_i , \\
\# \text{ vertices of degree three} &= 2h - 2 ,
\end{aligned}
$$

     # internal vertices of degree two $= 0$,
     # external vertices of degree two $= 2h + 4 - n_i$,
     # internal vertices of degree three $= n_i$,
     # external vertices of degree three $= 2h - 2 - n_i$,
     # internal edges $= h - 1 + n_i$,
     # external edges $= 4h + 2 - 2n_i$.

We say that an edge is an (α, β)-edge if it connects a vertex of degree α with a vertex of degree β (or vice versa). In benzenoid systems only $\alpha, \beta = 2, 3$ may occur. In this notation a free edge (see above) is just a $(2, 2)$-edge. One has:

     # $(2, 2)$-edges $= 6 + b$,
     # $(2, 3)$-edges $= 4h - 4 - 2b - 2n_i$,
     # $(2, 3)$-edges $= h - 1 + b + n_i$,

where b is a parameter usually called the number of bay regions (BALASUBRAMANIAN et al. 1980, KNOP et al. 1983). In fact,

     $b = $ # bays $+ 2 ($# coves$) + 3 ($# fjords$)$

or, simply,

     $b = $ # external $(3, 3)$-edges .

Furthermore, we have:

     # internal $(2, 2)$-edges $= 0$,
     # external $(2, 2)$-edges $= $ # free edges $= 6 + b$,
     # internal $(2, 3)$-edges $= 0$,
     # external $(2, 3)$-edges $= 4h - 4 - 2b - 2n_i$,
     # internal $(3, 3)$-edges $= h - 1 + n_i$.

We close this set of equalities by observing that

     size of perimeter $= $ # external vertices
                      $= $ # external edges
                      $= 4h + 2 - 2n_i$.

    The reader may immediately observe that many of the above equations are interrelated and some of them are trivial. It would be a good exercise for the reader to check all the given formulas on an example, maybe on the benzenoid system from Fig. 3.1.

    It is easy to deduce any formula of the above kind. Therefore these formulas should not be associated with anybody's name. A few sources, however, need to be mentioned: POLANSKY and ROUVRAY (1976a), GUTMAN (1985), DIAS (1987).

    HARARY and HARBORTH (1976) found additional relations between the numbers

n, h and m. Let $\{x\}$ denote the smallest integer which is greater than or equal to x. (For example, $\{6.9\} = 7$, $\{7\} = 7$, $\{7.1\} = 8$, $\{\pi\} = 4$.) Then for a benzenoid system with h hexagons, n vertices, and m edges,

$$2h + 1 + \{\sqrt{12h - 3}\} \le n \le 4h + 2,$$

$$3h + \{\sqrt{12h - 3}\} \le m \le 5h + 1,$$

$$\left\{\frac{1}{4}(n - 2)\right\} \le h \le n + 1 - \left\{\frac{1}{2}(n + \sqrt{6n})\right\},$$

$$n - 1 + \left\{\frac{1}{4}(n - 2)\right\} \le m \le 2n - \left\{\frac{1}{2}(n + \sqrt{6n})\right\},$$

$$\left\{\frac{1}{5}(m - 1)\right\} \le h \le m - \left\{\frac{1}{3}(2m - 2 + \sqrt{4m + 1})\right\},$$

$$1 + \left\{\frac{1}{3}(2m - 2 + \sqrt{4m + 1})\right\} \le n \le m + 1 - \left\{\frac{1}{5}(m - 1)\right\}.$$

These bounds are the best possible: any combination of h, n, and m satisfying the above inequalities may occur. In particular, benzenoid systems exist for all n, except $n < 6$ and $n = 7, 8, 9, 11, 12, 15$, and for all m, except $m < 6$ and $m = 7, 8, 9, 10, 12, 13, 14, 17, 18, 22$.

First classification: Catacondensed and pericondensed benzenoid systems

   A benzenoid system is said to be *catacondensed* if it does not possess internal vertices ($n_i = 0$). Otherwise, if $n_i > 0$, the benzenoid system is *pericondensed*. The pericondensed systems can be further divided according to the number of internal vertices.

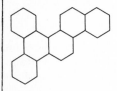   

$n_i = 0$ $n_i = 2$ $n_i = 4$

catacondensed   pericondensed benzenoid systems
benzenoid system

Any hexagon of a catacondensed benzenoid system is of the mode L_1, L_2, A_2, or A_3. If A_3-modes are absent, then the catacondensed system is *unbranched*. Otherwise it is *branched*.

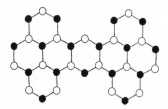

unbranched catacondensed
benzenoid system
(no A_3-modes)

branched catacondensed
benzenoid system

3.3 Coloring of Vertices

The vertices of a benzenoid system can be divided into two groups, such that vertices from the same group are never adjacent. It is customary to speak about *coloring of vertices*, which means that the vertices are colored by two colors (say black and white), so that adjacent vertices never have the same color.

Colored vertices in a benzenoid system. Observe that all vertices lying on a horizontal line have the same color. Here, $\Delta = 0$

It is easy to verify that if the color of one vertex is chosen, the colors of all other vertices are determined. By convention we color the vertices of the type (*a*) black and those of the type (*b*) white:

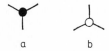

a   b

GORDON and DAVISON (1952) named the above vertex types *female* (*a*) and *male* (*b*), but this terminology is only rarely used.

The *color excess* of a benzenoid system is defined as

$$\Delta = |\# \text{ black vertices } - \# \text{ white vertices}| .$$

The examples below indicate that the color excess may assume any positive integer value (GUTMAN 1974).

$\Delta = 1$ $\Delta = 2$ $\Delta = 3$

For the majority of benzenoid systems of chemical relevance, the color excess is equal to zero; the reason for this is explained in Chapter 5.

3.4 Peaks and Valleys

Peaks and valleys are special vertices of degree two (belonging thus to the perimeter). A *peak* lies above both its first neighbors, a *valley* lies below both its first neighbors. In the two isomorphic benzenoid systems below, the peaks are marked white and the valleys black.

According to the coloring convention we have just adopted, the peaks are always colored white, whereas the valleys are always black.

From the above example it is seen that the number of peaks and valleys may change upon rotation of the benzenoid system. Their difference, however, remains invariant. It can be shown (CYVIN and GUTMAN 1987) that

$$\Delta = |\# \text{ peaks} - \# \text{ valleys}|,$$

which also provides us with an easy method of determining the color excess of a benzenoid system.

A *monotonic path* is a path connecting a peak with a valley in which, when starting at the peak, one always goes downwards. Two monotonic paths are indicated below.

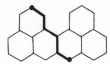 

3.5 1-Factors and KEKULÉ Structures

The correspondence between a KEKULÉ structure of a benzenoid hydrocarbon and a 1-factor of a benzenoid system has already been mentioned in the Interlude. We now provide a rigorous definition of a 1-factor.

Consider a benzenoid system B with n vertices and let n be an even number. A *1-factor* of B is a selection of $n/2$ edges in B, such that no two of them are incident. Hence, every vertex of B must be incident to one (and only one) of the selected edges.

In the example below, the selected edges are marked by heavy lines.

two 1-factors of benzo[a]pyrene

Another example is found in the Interlude.

In order to remain as close to chemistry as possible, we shall usually call a 1-factor a KEKULÉ *structure*. The selected edges will be called *double bonds* (in that KEKULÉ structure/1-factor). The edges which are not selected are *single bonds* (in that KEKULÉ structure/1-factor).

An edge of a benzenoid system is called a *fixed single bond* if it is not selected in any KEKULÉ structure. An edge is a *fixed double bond* if it is selected in all KEKULÉ structures.

Second classification (neo): Normal (n), essentially disconnected (e) and non-Kekuléan (o) benzenoid systems

Benzenoid systems possessing KEKULÉ structures are *Kekuléan*. Those not possessing KEKULÉ structures are *non-Kekuléan*. Kekuléan benzenoid systems are further divided into those which are *essentially disconnected* (having fixed double and/or single bonds) and those which are *normal* (having no fixed bonds). Non-Kekuléans are divided into *obvious non-Kekuléans* (if their color excess Δ is not zero) and *concealed non-Kekuléans* (if $\Delta = 0$).

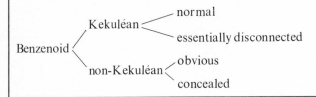

3.6 The Dualist Graph

SMITH (1961) proposed a very efficient shorthand representation of a benzenoid system, which is identical to the *dualist graph* (or, as it is sometimes called, the characteristic graph) advocated by BALABAN and HARARY (1968). The dualist graph is obtained when a vertex (i.e. a distinguished point) is drawn in the center of each hexagon, and vertices lying in adjacent hexagons are connected by straight lines.

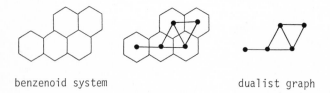

benzenoid system   dualist graph

Note that the angles between incident edges in a dualist graph must remain fixed.

Evidently, a benzenoid system can be immediately reconstructed from its dualist graph. Hence, there is a one-to-one correspondence between a benzenoid system and its dualist graph, and both have exactly the same combinatorial contents. Nevertheless, in certain cases it is more expedient to deal with dualist graphs than with benzenoid systems (BALABAN 1982). It is noted that the symmetry of a benzenoid system is maintained in its dualist graph:

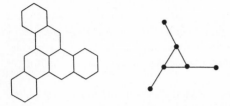

Both the benzenoid system and its dualist graph have C_{3h} symmetry

The linear acenes including benzene are exceptions. For them the dualist graph has higher symmetry than the corresponding benzenoid system.

Dualist graphs are not graphs in a mathematical sense of the word. The graph-theoretical counterpart of a dualist graph is the *inner dual*. The inner dual contains the same vertices and edges as the dualist graph, but now the length of the edges and the angles between them are immaterial. As a consequence of this, the inner dual does not characterize a benzenoid system up to isomorphism:

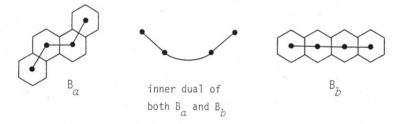

$$B_a \qquad \text{inner dual of} \qquad B_b$$
$$\text{both } B_a \text{ and } B_b$$

Nevertheless, the information contained in the inner dual may be sufficient in certain cases. For example, an inner dual is acyclic if and only if the corresponding benzenoid system is catacondensed. A less obvious fact is that the number of spanning trees of a benzenoid system can be deduced from its inner dual (GUT-MAN, MALLION and ESSAM 1983). (A spanning tree is a connected acyclic graph obtained by deleting some of the edges of a benzenoid graph.)

3.7 More about Symmetry

In every object which belongs to a symmetry group there is at least one point which remains unaffected by any of the symmetry operations. (This is the reason why symmetry groups are also called point groups.) For most of the groups of relevance to benzenoids, viz. D_{6h}, C_{6h}, D_{3h}, C_{3h}, D_{2h}, and C_{2h} (but not C_{2v} and C_s) a unique point remains unaffected by all the symmetry operations and is called the *center*. (For $\mathbf{C}_{2v}$ the points on a unique axis are unaffected, while all points are unaffected in the case of $\mathbf{C}_s$.) Some more detailed symmetry properties and subdivisions of the benzenoids are summarized in the following.

1. In benzenoids of the symmetries $\mathbf{D}_{6h}$ and $\mathbf{C}_{6h}$ the center always coincides with the center of a hexagon. This hexagon is called the *central hexagon*.

2. Among the $\mathbf{D}_{3h}$ and $\mathbf{C}_{3h}$ benzenoids, two kinds are distinguished: (i) the system has a central hexagon; (ii) the center coincides with a vertex, which is referred to as the *central vertex*.

The $\mathbf{D}_{3h}$ systems have three two-fold symmetry axes in the horizontal plane. (Also, vertical mirror planes pass through these axes.) All these three axes either (a) bisect a set of edges each, or (b) pass through vertices and edges. In the former case (a) one of the symmetry axes is horizontal with reference to the conventional orientation of the benzenoid, whereas in the latter case (b) one of the axes is vertical.

The $\mathbf{D}_{3h}$ benzenoid systems of the first kind (i) may have both types of two-fold symmetry axes described as (a) and (b). Hence, we subdivide these systems into (ia) and (ib) accordingly. The $\mathbf{D}_{3h}$ systems of the second kind (ii) are not divided further in this sense since all of the two-fold axes are of type (b). Below is a schematic survey.

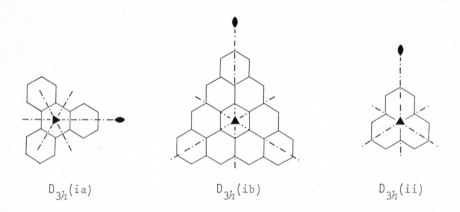

$D_{3h}(\text{ia})$ $D_{3h}(\text{ib})$ $D_{3h}(\text{ii})$

3. Benzenoids of $\mathbf{D}_{2h}$ and $\mathbf{C}_{2h}$ symmetries fall into two categories: (i) the system has a central hexagon; (ii) the center bisects an edge which is called the *central edge*.

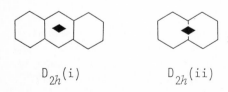

$D_{2h}(\text{i})$ $D_{2h}(\text{ii})$

4. A benzenoid of C_{2v} symmetry has no center, but a unique symmetry axis which is two-fold. (Also, a vertical mirror plane passes through this axis.) This axis either bisects edges (a), or passes through vertices and edges (b). The benzenoid may always be oriented (within the usual conventions) so that this axis is (a) horizontal or (b) vertical.

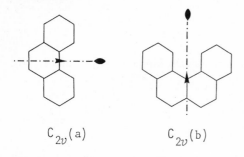

$C_{2v}(a)$ $C_{2v}(b)$

Third classification: Symmetry

1. *Hexagonal*, either D_{6h} (*regular hexagonal*) or C_{6h}.
2. *Trigonal*, either D_{3h} (*regular trigonal*) or C_{3h}. Two kinds: (i) with a central hexagon or (ii) with a central vertex. The regular trigonal systems may have (a) a horizontal two-fold symmetry axis or (b) a vertical two-fold axis.
3. *Dihedral*, D_{2h} and *centrosymmetrical*, C_{2h}. Two kinds: (i) with a central hexagon or (ii) with a central edge.
4. *Mirror-symmetrical*, C_{2v}. These systems may always be oriented (as usual, with two edges of each hexagon vertical) so that either (a) the symmetry axis is horizontal or (b) the symmetry axis is vertical.
5. *Unsymmetrical*, C_s.

Schematic survey:

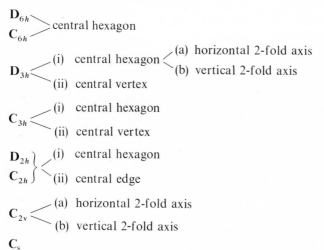

D_{6h}
C_{6h} } central hexagon

D_{3h} < (i) central hexagon < (a) horizontal 2-fold axis / (b) vertical 2-fold axis
            (ii) central vertex

C_{3h} < (i) central hexagon
            (ii) central vertex

D_{2h} }
C_{2h} } < (i) central hexagon
              (ii) central edge

C_{2v} < (a) horizontal 2-fold axis
            (b) vertical 2-fold axis

C_s

There are restrictions on the numbers of hexagons for benzenoids of special symmetries. Also, the color excess (Δ) has "allowed" and "forbidden" values under different symmetry conditions.

1. Benzenoids of hexagonal symmetry $\mathbf{D}_{6h}$ and $\mathbf{C}_{6h}$ occur for

$$h = 6\eta + 1; \qquad \eta = 0, 1, 2, \ldots ,$$

but $\eta > 2$ for $\mathbf{C}_{6h}$.

The only allowed Δ value is zero.

2. Benzenoids of trigonal symmetry of the first kind, viz. $\mathbf{D}_{3h}$ (i) and $\mathbf{C}_{3h}$ (i), have

$$h = 3\xi + 1; \qquad \xi = 1, 2, 3, \ldots ,$$

but not $\xi = 1$ for $\mathbf{C}_{3h}$ (i).

The allowed Δ values for the $\mathbf{D}_{3h}$ (i) and $\mathbf{C}_{3h}$ (i) systems are $\Delta = 3\eta, \eta = 0, 1, 2, \ldots$. Specifically, for the regular triangular systems, those of $\mathbf{D}_{3h}$ (ia) can only assume the values $\Delta = 0$. Systems of $\mathbf{D}_{3h}$ (ib) exist with any of the Δ values given above.

Benzenoids belonging to $\mathbf{D}_{3h}$ (ii) and $\mathbf{C}_{3h}$ (ii) have

$$h = 3\xi; \qquad \xi = 1, 2, 3, \ldots ,$$

but not $\xi = 1$ for $\mathbf{C}_{3h}$ (ii).

The allowed values for the color excess are exactly those which are forbidden for the benzenoids of trigonal symmetry of the first kind. They are $\Delta = 3\eta + 1$, $\Delta = 3\eta + 2; \eta = 0, 1, 2, \ldots$.

3. Dihedral (symmetry $\mathbf{D}_{2h}$) and centrosymmetrical ($\mathbf{C}_{2h}$) benzenoids of the first kind (i) occur for odd h values

$$h = 2\xi + 1; \qquad \xi = 1, 2, 3, \ldots ,$$

but not $\xi = 1$ for $\mathbf{C}_{2h}$ (i).

Benzenoids belonging to $\mathbf{D}_{2h}$ (ii) and $\mathbf{C}_{2h}$ (ii) have an even number of hexagons:

$$h = 2\xi; \qquad \xi = 1, 2, 3, \ldots ,$$

but not $\xi = 1$ for $\mathbf{C}_{2h}$ (ii).

In all cases of dihedral and centrosymmetrical benzenoids the only allowed value for the color excess is zero.

4. Mirror-symmetrical (symmetry $\mathbf{C}_{2v}$) benzenoids occur for all $h \geq 3$, but not $h = 3$ for $\mathbf{C}_{2v}(b)$, and not $h = 4$ for $\mathbf{C}_{2v}(a)$.

The $\mathbf{C}_{2v}$ (a) systems are restricted to $\Delta = 0$, while all Δ values are allowed for $\mathbf{C}_{2v}$ (b).

5. Unsymmetrical ($\mathbf{C}_s$) benzenoids are found for all $h \geq 4$. All Δ values are allowed for $\mathbf{C}_s$.

Chapter 4

Enumeration

By enumeration of benzenoid systems, the counting of all possible non-isomorphic members within a class of benzenoids is understood. Usually, but not always, the number of hexagons (h) is the leading parameter. Thus the enumeration for $h = 1, 2,$ 3, 4, etc. is to be executed. Figure 2.1 shows the results of enumerations for $h = 2, 3,$ and 4, which account for all possible benzenoid systems with these numbers of hexagons. For larger h values, it is of interest to subdivide the total amount of benzenoids into different classes. In this connection we speak about classification.

The enumeration of benzenoid systems may be traced back to KLARNER (1965), who followed up a suggestion of GOLOMB (1954). KLARNER (1965) gave numbers for systems with 1, 2, 3, 4, 5, and 6 hexagons. These works, among others (e.g. HARARY 1967, HARARY and READ 1970, LUNNON 1972), viewed the enumeration problems in a purely mathematical way. The first enumeration of benzenoids in a chemical context seems to be contained in the paper of BALABAN and HARARY (1968). After a period of some years with apparently no activity in this area the problems were taken up again, in pace with the access to modern computers (BALASUBRAMANIAN et al. 1980, KNOP et al. 1983). Soon this field really flared up. Twenty-one relevant publications from 1984 or later are cited in a consolidated report by fourteen authors (BALABAN et al. 1987). This report summarizes the data from literature which were available at the time it was written and gives also a number of supplements. The numbers of benzenoids up to $h = 11$ are contained therein. A substantial number of yet more recent publications exists. In particular, the enumeration of all benzenoids with $h = 12$ has been achieved (HE WJ et al. 1988, CIOSLOWSKI 1988) and also of the catacondensed systems with $h = 13$ (CIOSLOWSKI 1988). Some more of the recent publications in this field are cited in the subsequent sections of this chapter.

4.1 General Principles

Most of the algorithms for enumeration of benzenoid systems are based on the addition of hexagons in accord with Definition B of Chapter 2. Here the reader should especially recall the five types of additions, (i)–(v), and the fact that all benzenoid systems with $h + 1$ hexagons are generated by these additions to the set of benzenoids with h hexagons.

A crude generation of benzenoid systems by successive addition of hexagons inevitably leads to the production of identical or isomorphic systems; cf. Definition D of Chapter 2. Such duplicated systems are often produced in copious amounts, and a computer program must handle the elimination of duplicated systems in one way or another. A procedure to this effect is often the most time-consuming part of the program.

When it comes to a computer-aided classification of benzenoid systems, different cases or principles are distinguished (BRUNVOLL, CYVIN, and CYVIN 1987b):

(a) specific generation,
(b) recognition,
(c) exclusion.

Suppose that a broad class of benzenoid systems has been generated, e.g. all benzenoids with a given h. We speak about *recognition* when a special program or procedure in the main program detects the members of a subclass within the whole set. It may be more efficient, however, if it is possible to generate separately a smaller class of benzenoids within those with a given h. Then we speak about the principle of *specific generation*. The principle of *exclusion* is applicable when a class of benzenoids in one way or another has been divided into subclasses, but so that one subclass remains. Then the number of benzenoids in the last subclass is obviously obtained by subtraction.

Several approaches lrave been employed that cannot easily be characterized under the labels (a), (b), and (c) above. Let us use an example and describe the way it was proved for the first time (by computer-enumeration) that there are exactly 8 concealed non-Kekuléan benzenoid systems with 11 hexagons (BRUNVOLL, CYVIN SJ et al. 1987). (For definitions of the different classes, here and in the remainder of this chapter, the reader should consult Chapter 3.)

A concealed non-Kekuléan benzenoid system is recognized by the condition $\Delta = 0$, $K = 0$, a test which was incorporated into the enumeration program. It would be a pure application of the principle of recognition if the mentioned criterion was applied to all the existing 141 229 systems with $h = 11$. This number can be somewhat reduced by means of certain specific generations, but that is not the point of the present argument. We wish to describe a technique of successive elimination. During the generation of $h = 11$ benzenoids, every system was tested for the condition $\Delta = 0$, $K = 0$. When the condition was not fulfilled, as in the vast majority of cases, the system was immediately discarded. Only the systems with $\Delta = 0$, $K = 0$ were retained. In this way the testing for isomorphic systems was reduced to a minimum.

4.2 Symmetry

In Section 3.1 the eight possible symmetry groups for benzenoid systems are specified. It has almost become a standard procedure to build into a computer program a recognition of the symmetry group of a generated benzenoid system. Also, a number of specific generations of systems belonging to some of the symmetry groups have been executed. Results from these enumerations are collected in Table 4.1. This distribution of the benzenoids into different symmetry groups is only a coarse classification. We shall return to the issue of symmetry in a subsequent section, where the results from finer classifications are reported.

Table 4.1. Numbers of benzenoid systems belonging to the different symmetry groups (see also Table 4.15)

h	$\mathbf{D}_{6h}$	$\mathbf{C}_{6h}$	$\mathbf{D}_{3h}$	$\mathbf{C}_{3h}$	$\mathbf{D}_{2h}$	$\mathbf{C}_{2h}$	$\mathbf{C}_{2v}$	$\mathbf{C}_{s}$
1	1	0	0	0	0	0	0	0
2	0	0	0	0	1	0	0	0
3	0	0	1	0	1	0	1	0
4	0	0	1	0	2	1	1	2
5	0	0	0	0	2	1	9	10
6	0	0	1	1	3	7	12	57
7	1	0	1	1	3	7	39	279
8	0	0	0	0	6	35	61	1333
9	0	0	1	5	7	36	178	6278
10	0	0	4	5	11	169	274	29623
11	0	0	0	0	14	177	796	140242
12	0	0	3	21	21	807	*	*
13	2	0	4	26	23	859	*	*
14	0	0	0	0	41	*	*	*
15	0	0	3	95	50	*	*	*
16	0	0	12	118	80	*	*	*
17	0	0	0	0	94	*	*	*
18	0	0	6	423	156	*	*	*
19	2	2	19	543	189	*	*	*
20	0	0	0	0	310	*	*	*

* Unknown

Table 4.2. Numbers of benzenoid systems according to the first classification

h	Catacondensed	Pericondensed	Total
1	1	0	1
2	1	0	1
3	2	1	3
4	5	2	7
5	12	10	22
6	36	45	81
7	118	213	331
8	441	1024	1435
9	1489	5016	6505
10	5572	24514	30086
11	21115	120114	141229
12	81121	588463	669584
13	(315075)*	(2883181)*	3198256**
14	*	*	15367577**
15	*	*	74207930**

* Unknown. The parenthesized numbers are uncertain.
** Trinajstić N, Nikolić S, Knop JV, Müller WR, Szymanski K (1989) unpublished work.

4.3 First and Second Classification

In Chapter 3 the classification into catacondensed and pericondensed benzenoid systems was referred to as the first classification. The second classification was termed *neo*, referring to the normal (*n*), essentially disconnected (*e*), and non-Kekuléan (*o*) benzenoids. The results of enumeration in the frames of these classifications are given in Tables 4.2 and 4.3, respectively.

Table 4.3. Numbers of benzenoid systems according to the second classification (*neo*)*

h	*n*	*e*	*o*
1	1	0	0
2	1	0	0
3	2	0	1
4	6	0	1
5	14	1	7
6	48	3	30
7	167	23	141
8	643	121	671
9	2531	692	3282
10	10375	3732	15979
11	42919	19960	78350
12	**	**	384666

* *n* = normal; *e* = essentially disconnected; *o* = non-Kekuléan; the sum (*n* + *e* + *o*) equals the Total as given in Table 4.2.
** Unknown

4.4 Catacondensed Benzenoid Systems

The catacondensed benzenoid systems may be obtained by specific generation; cf. (a) of Section 4.1. This is achieved by successive additions starting with benzene, where the additions are restricted to one-contact additions only. A catacondensed benzenoid system may be unbranched or branched. These subclasses also may be generated specifically.

The unbranched systems are obtained by one-contact additions to the ends of already generated members. It is presupposed that the computer program can identify the end hexagons, which is not a big problem. It is achieved, for instance, by keeping track of the last added hexagon for both ends.

The branched systems are obtained by successive one-contact additions to all free edges of the already generated members, starting with triphenylene. This is the smallest catacondensed branched system.

Tables 4.4 and 4.5 list the numbers of unbranched (BALABAN et al. 1987, TOŠIĆ and KOVAČEVIĆ 1988) and branched (BALABAN et al. 1988) catacondensed systems, respectively. The distributions into symmetry groups are specified.

Table 4.4. Numbers of unbranched (*ub*) catacondensed benzenoid systems classified according to symmetry

h	$\mathbf{D}_{6h}$	$\mathbf{D}_{2h}$	$\mathbf{C}_{2h}$	$\mathbf{C}_{2v}$	$\mathbf{C}_s$	Total *ub*
1	1	0	0	0	0	1
2	0	1	0	0	0	1
3	0	1	0	1	0	2
4	0	1	1	1	1	4
5	0	1	1	4	4	10
6	0	1	4	3	16	24
7	0	1	4	12	50	67
8	0	1	13	10	158	182
9	0	1	13	34	472	520
10	0	1	39	28	1406	1474
11	0	1	39	97	4111	4248
12	0	1	116	81	11998	12196
13	0	1	115	271	34781	35168
14	0	1	339	226	100660	101226
15	0	1	336	764	290464	291565
16	0	1	988	638	837137	838764
17	0	1	977	2141	2408914	2412033
18	0	1	2866	1787	6925100	6929754
19	0	1	2832	6025	19888057	19896915
20	0	1	8298	5030	57071610	57084939

Table 4.5. Numbers of branched (*br*) catacondensed benzenoid systems classified according to symmetry

h	$\mathbf{D}_{3h}$	$\mathbf{C}_{3h}$	$\mathbf{D}_{2h}$	$\mathbf{C}_{2h}$	$\mathbf{C}_{2v}$	$\mathbf{C}_s$	Total *br*
4	1	0	0	0	0	0	1
5	0	0	0	0	1	1	2
6	0	0	1	0	4	7	12
7	1	1	1	0	4	44	51
8	0	0	1	4	18	206	229
9	0	0	1	4	27	937	969
10	2	4	3	25	67	3997	4098
11	0	0	4	26	118	16719	16867
12	0	0	4	132	269	68520	68925
13	2	15	*	*	*	*	*
14	0	0	*	*	*	*	*
15	0	0	*	*	*	*	*

* Unknown

4.5 Periodensed Benzenoid Systems

The periodensed benzenoid systems also may be obtained by specific generation. One starts with phenalene, the smallest periodensed system, and executes the additions of all kinds: one-, two-, three-, four-, and five-contact additions.

It is more interesting that the classes of normal and normal pericondensed benzenoid systems also seem to be available from specific generations. The principles are based on a conjecture that says any normal benzenoid with $h + 1$ hexagons may be generated by adding one hexagon to a normal benzenoid with h hexagons (CYVIN and GUTMAN 1986a)[*]. In this process only the one-, three- and five-contact additions are operating. These three types of additions are referred to as *normal additions*.

Suppose that all normal additions are executed successively. Then, if the process is started from benzene, all the normal benzenoid systems are obtained. They comprise both the catacondensed and pericondensed systems. The normal pericondensed systems are generated specifically if the mentioned process is started from pyrene, which is the smallest normal pericondensed benzenoid system. Results of enumerations for the normal pericondensed benzenoid systems are collected in Table 4.6. Those for $h \leq 10$ are taken from BRUNVOLL, CYVIN, and CYVIN (1987b). Enumerations of benzenoid systems with hexagonal symmetries (BRUNVOLL, CYVIN, and CYVIN 1987b, CYVIN, BRUNVOLL, and CYVIN 1989) and trigonal symmetries (CYVIN, BRUNVOLL, and CYVIN 1988a) have been treated in particular. The numbers of "Total np" (Table 4.6) plus the numbers of "Catacondensed" (Table 4.2) are consistent with the numbers of total normal benzenoid systems ("n" in Table 4.3).

Table 4.6. Numbers of normal pericondensed (np) benzenoid systems classified according to symmetry (see also Table 4.15)

h	D_{6h}	C_{6h}	D_{3h}	C_{3h}	D_{2h}	C_{2h}	C_{2v}	C_s	Total np
4	0	0	0	0	1	0	0	0	1
5	0	0	0	0	0	0	1	1	2
6	0	0	0	0	1	2	3	6	12
7	1	0	0	0	1	0	6	41	49
8	0	0	0	0	2	11	19	200	232
9	0	0	0	0	3	3	39	997	1042
10	0	0	1	0	6	52	90	4654	4803
11	0	0	0	0	*	*	*	*	21804
12	0	0	0	0	*	*	*	*	*
13	2	0	2	3	*	*	*	*	*

* Unknown

All essentially disconnected benzenoid systems are pericondensed. The numbers of such systems have so far (BRUNVOLL, CYVIN, et al. 1988a) been determined by the principle of exclusion; cf. (c) in Section 4.1. Numerical data are collected in Table 4.7. The total numbers are found under "e" in Table 4.3.

There is, however, a subtle method available for recognition, in the sense (b) of Section 4.1, of an essentially disconnected system. It is characterized by possessing bonds with PAULING bond order (cf. Chapter 5) equal to zero. The PAULING bond orders for benzenoid systems are known to be equal to the elements of the

---

[*] Recently this conjecture was proved to be correct by HE WC and HE WJ (unpublished work 1988).

Table 4.7. Numbers of essentially disconnected (*e*) benzenoid systems classified according to symmetry (see also Table 4.15)

h	$\mathbf{D}_{6h}$	$\mathbf{C}_{6h}$	$\mathbf{D}_{3h}$	$\mathbf{C}_{3h}$	$\mathbf{D}_{2h}$	$\mathbf{C}_{2h}$	$\mathbf{C}_{2v}$	$\mathbf{C}_s$
5	0	0	0	0	1	0	0	0
6	0	0	0	0	0	1	0	2
7	0	0	0	0	0	3	6	14
8	0	0	0	0	2	7	2	110
9	0	0	0	0	2	16	29	645
10	0	0	0	0	1	53	31	3 647
11	0	0	0	0	2	87	166	19 705
12	0	0	0	0	5	*	*	*
13	0	0	0	0	7	*	*	*
14	0	0	0	0	7	*	*	*
15	0	0	0	0	9	*	*	*
16	0	0	1	2	19	*	*	*
17	0	0	0	0	27	*	*	*
18	0	0	0	0	34	*	*	*
19	0	0	0	23	39	*	*	*
20	0	0	0	0	84	*	*	*

* Unknown

inverse adjacency matrix, (5.19). Therefore, a formulation of the necessary and sufficient conditions for a Kekuléan benzenoid system to be essentially disconnected, convenient for computer implementation, reads: $(A)_{ij} = 1, (A^{-1})_{ij} = 0$ for at least one pair of vertices, *ij*. Here, A is the adjacency matrix.

Finally, we have the non-Kekuléan systems among the pericondensed benzenoids. When a computer program has the computation of the color excess (Δ, see Chapter 3) built in, then the obvious non-Kekuléans are easily recognized by means of $\Delta > 0$. Computerized procedures for obtaining the number of KEKULÉ structures (K) are also available. Non-Kekuléan benzenoids have $K = 0$. If, in

Table 4.8. Numbers of obvious non-Kekuléan (*oo*) benzenoid systems classified according to symmetry (see also Table 4.16)

h	$\mathbf{D}_{3h}$	$\mathbf{C}_{3h}$	$\mathbf{C}_{2v}$	$\mathbf{C}_s$	Total *oo*
3	1	0	0	0	1
4	0	0	0	1	1
5	0	0	3	4	7
6	1	1	2	26	30
7	0	0	11	130	141
8	0	0	12	659	671
9	1	5	49	3 227	3 282
10	1	1	58	15 919	15 979
11	0	0	221	78 121	78 342
12	3	21	*	*	384 568

* Unknown

addition, $\Delta = 0$, then the system examined is concealed non-Kekuléan. Results of the enumeration of obvious (GUTMAN and CYVIN 1988a) and concealed (BRUN-VOLL, CYVIN SJ, et al. 1987, HE WC et al. 1988, CYVIN, BRUNVOLL, and CYVIN 1988b) benzenoid systems are collected in Tables 4.8 and 4.9, respectively. The total ($oo + co$) is consistent with Table 4.3.

Table 4.9. Numbers of concealed non-Kekuléan (co) benzenoid systems classified according to symmetry

h	$\mathbf{D}_{6h}$	$\mathbf{C}_{6h}$	$\mathbf{D}_{3h}$	$\mathbf{C}_{3h}$	$\mathbf{D}_{2h}$	$\mathbf{C}_{2h}$	$\mathbf{C}_{2v}$	$\mathbf{C}_{s}$	Total co
11	0	0	0	0	1	2	1	4	8
12	0	0	0	0	0	5	0	93	98

4.6 Classification According to the Number of Internal Vertices or the Number of External Vertices, and Enumeration of Chemical Isomers

An important classification of benzenoid systems follows their number of internal vertices. Then the class defined by $n_i = 0$ is coincident with the class of catacondensed benzenoid systems. Benzenoid systems with odd numbers n_i, viz. 1, 3, 5, ..., are necessarily obvious non-Kekuléans. Systems with $n_i = 2, 4, 6, ...$ may be either Kekuléan or non-Kekuléan.

Introduce the notation

$$\# \text{ external vertices} = n_e .$$

Then n_e is also the number of edges of the perimeter. One of the relations in Chapter 3 may now be written as

$$n_e = 4h + 2 - 2n_i .$$

Thus it is clear that the numbers of benzenoid systems with given h and n_i values also may be interpreted in terms of a classification according to the n_e values.

The values of both n_i and n_e for a given h are limited. Specifically, one has (GUTMAN 1982a, BALABAN et al. 1987)

$$0 \leq n_i \leq 2h + 1 - \{\sqrt{12h - 3}\} ,$$
$$2\{\sqrt{12h - 3}\} \leq n_e \leq 4h + 2 ,$$

where the meaning of the symbol $\{x\}$ is the same as in the Section 3.2; there it was used in connection with the HARARY-HARBORTH inequalities.

Table 4.10 shows the numbers of benzenoid systems with $h \leq 10$, classified according to the numbers of internal vertices n_i (KNOP et al. 1983). This classi-

Table 4.10. Numbers of benzenoid systems classified according to their number of internal vertices (n_i)

h	n_i										
	0	1	2	3	4	5	6	7	8	9	10
1	1	0	0	0	0	0	0	0	0	0	0
2	1	0	0	0	0	0	0	0	0	0	0
3	2	1	0	0	0	0	0	0	0	0	0
4	5	1	1	0	0	0	0	0	0	0	0
5	12	6	3	1	0	0	0	0	0	0	0
6	36	24	14	4	3	0	0	0	0	0	0
7	118	106	68	25	10	3	1	0	0	0	0
8	411	453	329	144	67	21	9	1	0	0	0
9	1489	1966	1601	825	396	154	55	15	4	0	0
10	5572	8395	7652	4518	2340	1018	416	123	42	9	1

fication is also known for $h = 11$ (STOJMENOVIĆ et al. 1986, BALABAN et al. 1987) and for $h = 12$ (HE WJ et al. 1988). The same numbers as in Table 4.19 are found in Table 4.11, where they represent the enumeration according to the perimeter length n_e. This classification is known completely for $n_e \leq 46$ (STOJMENOVIĆ et al. 1986). It is clear that the number of all benzenoid systems with a given n_e value is limited. Nonvanishing values occur for $n_e = 6$ and all even $n_e \geq 10$, see Table 4.12.

The enumeration of chemical isomers of benzenoid hydrocarbons (DIAS 1982), say $C_n H_s$, is a problem closely connected to those of the present section. The number of carbon atoms corresponds to the number of vertices, viz. n. The number of hydrogen atoms is

$$\# \text{ vertices of degree two} = s .$$

This corresponds to the number of secondary carbon atoms; hence, the present symbol s. In Chapter 3 the invariants n and s are expressed in terms of h and n_i. The corresponding relations in terms of h and n_e read:

$$n = \frac{1}{2}(4h + n_e + 2) ;$$
$$s = \frac{1}{2}(n_e + 6) .$$

It is interesting to notice that the last relation is independent of h. One has also:

$$s = n - 2h + 2 .$$

In accord with a statement in Chapter 3, the possible values of n are 6, 10, 13, 14, and $n \geq 16$. Correspondingly, it is found that $s = 6$ or $s \geq 8$. The upper and

Table 4.11. Numbers of benzenoid systems classified according to their number of external vertices or perimeter length (n_e)

									n_e									
h	6	10	12	14	16	18	20	22	24	26	28	30	32	34	38	40	42	44
1	1	0	0	0	0	0	0	0	0	0	0	0	0	0	0	0	0	0
2	0	1	0	0	0	0	0	0	0	0	0	0	0	0	0	0	0	0
3	0	0	1	2	0	0	0	0	0	0	0	0	0	0	0	0	0	0
4	0	0	0	1	1	5	0	0	0	0	0	0	0	0	0	0	0	0
5	0	0	0	0	1	3	6	12	0	0	0	0	0	0	0	0	0	0
6	0	0	0	0	0	3	4	14	24	36	0	0	0	0	0	0	0	0
7	0	0	0	0	0	1	3	10	25	68	106	118	0	0	0	0	0	0
8	0	0	0	0	0	0	1	9	21	67	144	329	453	411	0	0	0	0
9	0	0	0	0	0	0	0	4	15	55	154	396	825	1601	1966	1489	0	0
10	0	0	0	0	0	0	0	1	9	42	123	416	1018	2340	4518	7652	8395	5572

Table 4.12. Numbers of benzenoid systems with given n_e values

n_e	Number	n_e	Number
6	1	28	744
10	1	30	2291
12	1	32	6186
14	3	34	18714
16	2	36	53793
18	12	38	162565
20	14	40	482416
22	50	42	1467094
24	97	44	4436536
26	312	46	13594266

lower bounds of n for a given value of h are presented in Section 3.2. The analogous statements for s read:

$$3 + \{\sqrt{12h - 3}\} \le s \le 2h + 4 ,$$

$$2\left\{\frac{1}{2}\left(n + \sqrt{6n}\right)\right\} - n \le s \le n + 2 - 2\left\{\frac{1}{4}(n - 2)\right\} .$$

Now we are in a position to map all the types of chemical isomers of benzenoid hydrocarbons for increasing h values. The formulas for $h \le 10$ are shown below:

$h = 1$: C_6H_6

$h = 2$: $C_{10}H_8$

$h = 3$: $C_{13}H_9$, $C_{14}H_{10}$

$h = 4$: $C_{16}H_{10}$, $C_{17}H_{11}$, $C_{18}H_{12}$

$h = 5$: $C_{19}H_{11}$, $C_{20}H_{12}$, $C_{21}H_{13}$, $C_{22}H_{14}$

$h = 6$: $C_{22}H_{12}$, $C_{23}H_{13}$, $C_{24}H_{14}$, $C_{25}H_{15}$, $C_{26}H_{16}$

$h = 7$: $C_{24}H_{12}$, $C_{25}H_{13}$, $C_{26}H_{14}$, $C_{27}H_{15}$, $C_{28}H_{16}$, $C_{29}H_{17}$, $C_{30}H_{18}$

$h = 8$: $C_{27}H_{13}$, $C_{28}H_{14}$, $C_{29}H_{15}$, $C_{30}H_{16}$, $C_{31}H_{17}$, $C_{32}H_{18}$, $C_{33}H_{19}$, $C_{34}H_{20}$

$h = 9$: $C_{30}H_{14}$, $C_{31}H_{15}$, $C_{32}H_{16}$, $C_{33}H_{17}$, $C_{34}H_{18}$, $C_{35}H_{19}$, $C_{36}H_{20}$, $C_{37}H_{21}$, $C_{38}H_{22}$

$h = 10$: $C_{32}H_{14}$, $C_{33}H_{15}$, $C_{34}H_{16}$, $C_{35}H_{17}$, $C_{36}H_{18}$, $C_{37}H_{19}$, $C_{38}H_{20}$, $C_{39}H_{21}$, $C_{40}H_{22}$, $C_{41}H_{23}$, $C_{42}H_{24}$

The numbers of these isomers are collected in Table 4.13. These numbers are exactly the same as those of Tables 4.10 or 4.11.

DIAS (1982, 1987) has enumerated (without computer aid) some of the C_nH_s isomers. His numbers are not immediately comparable with those of Table 4.13 because he has counted only the Kekuléan systems. Furthermore, he has included helicenic systems, at least in one of the references (DIAS 1982).

Table 4.13. Numbers of C_nH_s isomers

						s						
n	6	8	9	10	11	12	13	14	15	16	17	18
6	1	0	0	0	0	0	0	0	0	0	0	0
10	0	1	0	0	0	0	0	0	0	0	0	0
13	0	0	1	0	0	0	0	0	0	0	0	0
14	0	0	0	2	0	0	0	0	0	0	0	0
16	0	0	0	1	0	0	0	0	0	0	0	0
17	0	0	0	0	1	0	0	0	0	0	0	0
18	0	0	0	0	0	5	0	0	0	0	0	0
19	0	0	0	0	1	0	0	0	0	0	0	0
20	0	0	0	0	0	3	0	0	0	0	0	0
21	0	0	0	0	0	0	6	0	0	0	0	0
22	0	0	0	0	0	3	0	12	0	0	0	0
23	0	0	0	0	0	0	4	0	0	0	0	0
24	0	0	0	0	0	1	0	14	0	0	0	0
25	0	0	0	0	0	0	3	0	24	0	0	0
26	0	0	0	0	0	0	0	10	0	36	0	0
27	0	0	0	0	0	0	1	0	25	0	0	0
28	0	0	0	0	0	0	0	9	0	68	0	0
29	0	0	0	0	0	0	0	0	21	0	106	0
30	0	0	0	0	0	0	0	4	0	67	0	118

						s					
n	14	15	16	17	18	19	20	21	22	23	24
31	0	15	0	144	0	0	0	0	0	0	0
32	1	0	55	0	329	0	0	0	0	0	0
33	0	9	0	154	0	453	0	0	0	0	0
34	0	0	42	0	396	0	411	0	0	0	0
35	0	*	0	123	0	825	0	0	0	0	0
36	0	0	*	0	416	0	1601	0	0	0	0
37	0	*	0	*	0	1018	0	1966	0	0	0
38	0	0	*	0	*	0	2340	0	1489	0	0
39	0	0	0	*	0	*	0	4518	0	0	0
40	0	0	*	0	*	0	*	0	7652	0	0
41	0	0	0	*	0	*	0	*	0	8395	0
42	0	0	*	0	*	0	*	0	*	0	5572

* Unknown

4.7 Classification According to Color Excess

The benzenoid systems with a given h are distributed over the Δ values (color excess) in the range

$$0 \le \Delta \le \left[\frac{1}{3}h\right],$$

where $[x]$ signifies the largest integer smaller than or equal to x. The upper and lower bounds of Δ are always realized (BRUNVOLL, CYVIN, et al. 1988b).

The properties of benzenoid systems with different color excess have been studied in particular by BRUNVOLL, CYVIN, et al. (1988b) in connection with computer-aided enumerations. Certain rules, referred to as *selection rules for Δ*, have been exploited in this connection.

Assume that a hexagon is added to a benzenoid system with parameters h and Δ. Let the corresponding parameters for the new benzenoid system be $h' = = h + 1$ and Δ'. The selection rules state that

$$\Delta' - \Delta = 0, \pm 1 \qquad (\text{not} -1 \text{ if } \Delta = 0) .$$

More specifically, if the addition is a one-, three- or five-contact addition (that is to say, normal addition), then the color excess does not change ($\Delta' = \Delta$). In the case of two- or four-contact additions, the color excess shifts by one unit ($\Delta' = \Delta \pm 1$).

Especially interesting are the extremal obvious non-Kekuléan systems defined by

$$h = 3\Delta, \qquad \Delta = \Delta_{max} = \frac{1}{3}h \qquad (\Delta > 0) .$$

These systems evidently occur for $h = 3, 6, 9, \dots$. It has been proved that they all consist of the so-called *teepees* (also *TP* benzenoids, where T and P stand for "Triangulene" and "Phenalene", respectively). A teepee is by definition phenalene ($h = 3$), triangulene ($h = 6$) or a benzenoid system in which only these two kinds of triangular units are fused (i.e. where two neighboring units share exactly one edge), and the triangle apex of each unit points the same way (conventionally, upwards). Below we show the three existing teepees with $h = 9, \Delta = 3$.

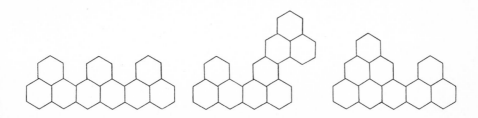

The systems with $\Delta = 0$ and $h \le 10$ comprise the catacondensed benzenoids (Tables 4.3 and 4.4), normal pericondensed and essentially disconnected benzenoids (Tables 4.5 and 4.6, respectively). For $h \ge 11$, the concealed non-Kekuléans (Table 4.8) come in addition. In Table 4.14 we give the numbers of obvious non-Kekuléan benzenoid systems with different values of the color excess ($\Delta > 0$). The sums under "Total" (Table 4.14) for each h are consistent with "Total oo" of Table 4.8.

The data of Table 4.14 are from GUTMAN and CYVIN (1988b).

Table 4.14. Numbers of obvious non-Kekuléan benzenoid systems with different Δ values classified according to symmetry

h	Δ	$\mathbf{D}_{3h}$	$\mathbf{C}_{3h}$	$\mathbf{C}_{2v}$	$\mathbf{C}_{s}$	Total
3	1	1	0	0	0	1
4	1	0	0	0	1	1
5	1	0	0	3	4	7
6	1	0	1	1	26	28
	2	1	0	1	0	
7	1	0	0	10	124	134
	2	0	0	1	6	7
8	1	0	0	5	614	619
	2	0	0	7	45	52
9	1	0	4	39	2914	2957
	2	1	1	9	311	322
	3	0	0	1	2	3
10	1	0	0	20	14004	14024
	2	0	0	38	1878	1916
	3	1	1	0	37	39
11	1	0	0	156	66890	67046
	2	0	0	52	10870	10922
	3	0	0	13	361	374
12	1	3	13	*		320859
	2	0	7	* 176	60522	60705
	3	0	0	5	2985	2990
	4	0	1	6	7	14
13	1	0	0	*	*	*
	2	0	0	*	*	*
	3	0	8	84	21583	21675
	4	0	0	4	207	211
14	1	0	0	*	*	*
	2	0	0	*	*	*
	3	0	0	*	*	*
	4	0	0	46	2542	2588
15	1	0	58	*	*	*
	2	2	29	*	*	*
	3	0	0	*	*	*
	4	1	7	*	*	*
	5	0	1	4	43	48

* Unknown

4.8 More about Symmetry

In Table 4.1 the distributions of benzenoid systems with given h values into the eight symmetry groups are specified. In the present section we take into account the subdivisions under different symmetry groups, as were explained in Section 3.7. Further, the *neo* classification is accounted for, and the color excess is indicated.

It is seen from Table 4.1 that the benzenoids with hexagonal symmetry ($\mathbf{D}_{6h}$ or $\mathbf{C}_{6h}$) are very sparsely distributed among the total amount. Fortunately, it is possible to generate these systems specifically by adding six hexagons at a time

Table 4.15. Numbers of benzenoid systems with hexagonal symmetry; all of them have a central hexagon and $\Delta = 0$

h	D_{6h}				C_{6h}			
	n	e	$n+e$	o	n	e	$n+e$	o
1	1	0	1	0	0	0	0	0
7	1	0	1	0	0	0	0	0
13	2	0	2	0	0	0	0	0
19	2	0	2	0	2	0	2	0
25	2	1	3	0	7	1	8	0
31	5	0	5	0	24	8	32	0
37	7	1	8	0	84	44	128	0
43	11	1	12	1	*	*	523	4
49	17	3	20	0	*	*	2167	42
55	30	4	34	1	*	*	9158	312
61	*	*	59	1	*	*	*	*
67	*	*	100	4	*	*	*	*
73	*	*	176	7	*	*	*	*

* Unknown

Table 4.16. Numbers of benzenoid systems with trigonal symmetry

h	kind	Δ	D_{3h}			C_{3h}		
			n	e	o	n	e	o
3	(ii)	1	—	—	1	—	—	0
4	(i)	0	1	0	0	0	0	0
6	(ii)	1	—	—	0	—	—	1
		2	—	—	1	—	—	0
7	(i)	0	1	0	0	1	0	0
9	(ii)	1	—	—	0	—	—	4
		2	—	—	1	—	—	1
10	(i)	0	3	0	0	4	0	0
		3	—	—	1	—	—	1
12	(ii)	1	—	—	3	—	—	13
		2	—	—	0	—	—	7
		4	—	—	0	—	—	1
13	(i)	0	4	0	0	18	0	0
		3	—	—	0	—	—	8
15	(ii)	1	—	—	0	—	—	58
		2	—	—	2	—	—	29
		4	—	—	1	—	—	7
		5	—	—	0	—	—	1
16	(i)	0	10	1	0	73	2	0
		3	—	—	1	—	—	43
18	(ii)	1	—	—	3	—	—	234
		2	—	—	2	—	—	136
		4	—	—	0	—	—	45
		5	—	—	1	—	—	8
19	(i)	0	17	0	0	298	23	0
		3	—	—	2	—	—	217
		6	—	—	0	—	—	5

(BRUNVOLL, CYVIN, and CYVIN 1987b). The regular hexagonal ($\mathbf{D}_{6h}$) systems alone may be generated specifically by adding six or twelve hexagons at a time. Results of enumeration (BRUNVOLL, CYVIN, and CYVIN 1987b, HE WC et al. 1988, CYVIN, BRUNVOLL, and CYVIN 1988b, 1989) are collected in Table 4.15.

Table 4.16 summarizes the available enumeration data for benzenoid systems with trigonal symmetry ($\mathbf{D}_{3h}$ or $\mathbf{C}_{3h}$); cf. CYVIN, BRUNVOLL, and CYVIN (1988a). It is recalled that kinds (i) and (ii) refer to systems with central hexagon and central vertex, respectively.

4.9 All-Benzenoid Systems

The class of all-benzenoid systems is defined in Chapter 7. It is a subclass of the normal benzenoid systems. An all-benzenoid may be catacondensed, but only for every third value of h; specifically $h = 1, 4, 7, \dots$. Pericondensed all-benzenoids occur for $h = 6$ and $h \geq 8$. The all-benzenoid systems have been generated and enumerated by recognition up to $h = 10$ (KNOP et al. 1986a) and up to higher h

Table 4.17. Numbers of catacondensed all-benzenoid systems classified according to symmetry

h	$\mathbf{D}_{6h}$	$\mathbf{D}_{3h}$	$\mathbf{C}_{3h}$	$\mathbf{D}_{2h}$	$\mathbf{C}_{2h}$	$\mathbf{C}_{2v}$	$\mathbf{C}_{s}$	Total
1	1	0	0	0	0	0	0	1
4	0	1	0	0	0	0	0	1
7	0	0	0	1	0	1	0	2
10	0	1	0	0	0	2	3	6
13	0	1	1	0	3	7	20	32
16	0	0	0	0	0	14	158	172
19	0	0	2	1	19	41	1076	1139
22	0	0	8	0	0	79	7574	7661

Table 4.18. Numbers of periicondensed all-benzenoid systems classified according to symmetry

h	$\mathbf{D}_{6h}$	$\mathbf{C}_{6h}$	$\mathbf{D}_{3h}$	$\mathbf{C}_{3h}$	$\mathbf{D}_{2h}$	$\mathbf{C}_{2h}$	$\mathbf{C}_{2v}$	$\mathbf{C}_{s}$	Total
6	0	0	0	0	1	0	0	0	1
8	0	0	0	0	0	0	1	0	1
9	0	0	0	0	0	0	1	2	3
10	0	0	1	0	0	1	1	0	3
11	0	0	0	0	2	0	2	6	10
12	0	0	0	0	1	2	5	21	29
13	1	0	0	0	0	0	3	21	25
14	0	0	0	0	1	2	12	87	102
15	0	0	0	0	0	2	14	243	259
16	0	0	1	1	2	9	18	323	354
17	0	0	0	0	2	11	38	1085	1136
18	0	0	0	0	1	22	58	2632	2713

values by specific generation (CYVIN BN, BRUNVOLL, et al. 1988); see Tables 4.17 and 4.18.

In Table 4.17 there is no column for C_{6h} because there are no catacondensed all-benzenoids belonging to this symmetry. A unique catacondensed all-benzenoid system ($h = 1$, benzene) belongs to D_{6h}. In Table 4.18 all entries under C_{6h} are zero; the smallest all-benzenoid system with this symmetry has $h = 31$.

Chapter 5

Kekulé Structures

It is a wide-spread belief among chemists that KEKULÉ structures are used only within resonance theory (which, on the other hand, would be an oversimplified version of valence bond theory), and are thus of little or no interest for contemporary theoretical chemistry. In this chapter we try to convince the reader that in reality the situation is somewhat different. Recent work in *ab initio* valence bond theory (COOPER et al. 1986, GERRATT 1987) shows that the importance of KEKULÉ structures is much greater than is usually presumed, a fact which was known half a century ago, but eventually neglected by the broader scientific community (PAULING 1987).

In this book we pursue much simpler conceptual schemes than *ab initio* valence bond theory, where, nevertheless, KEKULÉ structures play a central role. Whether these "simple" theories have any relation to the valence bond theory or to some other sophisticated quantum chemical approach is not certain. What is certain, however, is that they provide a more or less direct insight into the electronic struc-

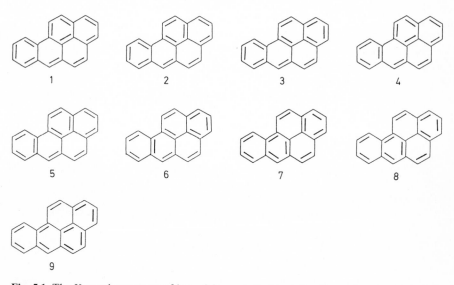

Fig. 5.1. The KEKULÉ structures of benzo[*a*]pyrene. By inspecting these nine structures, it is not easy to verify that they all are different. It is even less easy to see that benzo[*a*]pyrene has no more KEKULÉ structures, i.e. that $K = 9$

ture of large and very large benzenoid molecules, enable one to make predictions of their various properties on a semiquantitative level, and can be used (and even understood) by non-specialists, that is experimental organic chemists. In this and in the following two chapters we present several current chemical theories, all based on or related to Kekulé structures. To do this we have first to learn the basic facts about Kekulé structures and, in particular, their enumeration.

The relation between a Kekulé structure of a benzenoid hydrocarbon and a 1-factor of a benzenoid system has already been explained in the Interlude. At this point it would be useful for the reader to refresh his memory by reading again the few simple definitions concerning 1-factors and Kekulé structures that are given in Chapter 3.

Throughout this book the number of Kekulé structures is denoted by K. Instead of "number of Kekulé structures", one often says "Kekulé structure count" or just "K number". The most direct way to find out K is to draw all possible Kekulé structural formulas of a benzenoid molecule. An illustration of this is given in Fig. 5.1 where the Kekulé structures of benzo[a]pyrene are depicted.

The method of drawing the Kekulé structures is, obviously, hopelessly impractical for larger benzenoid systems in which hundreds or thousands of such structures may occur. Fortunately, there exist much more efficient ways to establish the value of K. Some of them are outlined in the next section.

5.1 Methods for Kekulé Structure Enumeration

In addition to direct drawing and counting, which as Fig. 5.1 shows is infeasible for larger benzenoids, there are several techniques for Kekulé structure enumeration. A recent book (Cyvin and Gutman 1988) is devoted to this problem, to which an interested reader is directed for further details and references. Here we briefly summarize the most powerful enumeration procedures.

5.1.1 Method of Fragmentation

The method of fragmentation is the oldest, simplest and most general enumeration scheme. It was already known in 1933 (Wheland 1933). It consists of two rules. Both rules apply to arbitrary conjugated molecules or even to arbitrary graphs. Here, the use of the graph representation of a benzenoid system is advantageous.

Rule 1. Let B be a (not necessarily) benzenoid graph and e its edge, connecting the vertices u and v. Then

$$K\{B\} = K\{B - e\} + K\{B - u - v\} , \tag{5.1}$$

where $B - e$ denotes the graph obtained from B by deleting the edge e; similarly, $B - u - v$ is obtained by deleting u and v (as well as all edges incident to them) from B.

Rule 2. If a graph B is composed of several mutually disconnected parts B_1, B_2, ... , B_p, then

$$K\{B\} = K\{B_1\} \, K\{B_2\} \, ... \, K\{B_p\} . \tag{5.2}$$

Rule 2 is usually employed after Rule 1, because $B - e$ and/or $B - u - v$ may contain several disconnected parts. A frequently used immediate consequence of Rules 1 and 2 is

Rule 3. If u is a vertex of degree one of the graph B, and v is its (unique) neighbor, then

$$K\{B\} = K\{B - u - v\} . \tag{5.3}$$

The fragments obtained by applying Rules 1–3 (when necessary, several times), are smaller and have a simpler structure than the molecular graph examined. So, by repeated application of Rules 1–3, we will ultimately come to fragments whose KEKULÉ structure counts are known. We illustrate this on the example of benzo[a]-pyrene (cf. Fig. 5.1). The edge where the system is "attacked" is indicated by an arrow.

First apply (5.1)

Now, knowing that the K values of benzene and naphthalene are 2 and 3, respectively, and that systems with odd numbers of vertices necessarily have $K = 0$, we further have:

Here (5.3) is applied four times in succession:

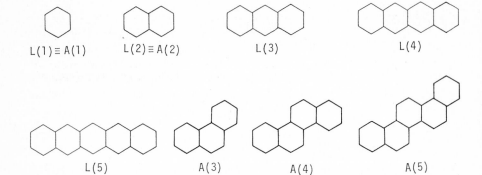

This finally gives

$$K\{\text{benzo}[a]\text{pyrene}\} = 6 + 3 = 9 .$$

5.1.2 Recurrence Relations

Denote by $L(h)$ and $A(h)$ the linear and the zig-zag benzenoid chains with h hexagons. The first members of these homologous series are:

L(1) ≡ A(1)   L(2) ≡ A(2)   L(3)   L(4)

L(5)   A(3)   A(4)   A(5)

By using Rules 1 and 3 we can easily see that

This means that the following relation is obeyed:

$$K\{L(h)\} = K\{L(h-1)\} + 1 \,. \tag{5.4}$$

In a similar manner:

This results in

$$K\{A(h)\} = K\{A(h-1)\} + K\{A(h-2)\} \,. \tag{5.5}$$

The equalities (5.4) and (5.5) are *recurrence relations*, because they enable one to calculate the K value of one member of the homologous series (i.e. $L(h)$ or $A(h)$) from the knowledge of the K values of the lower members of the same series (i.e. $L(h-1)$ or $A(h-1)$, $A(h-2)$). For any particular value of h, $K\{L(h)\}$ and $K\{A(h)\}$ can be evaluated step-by-step, starting with the *initial conditions* $K\{L(1)\} \equiv \equiv A(1)\} = 2$, $K\{L(2) \equiv A(2)\} = 3$. Hence,

$$K\{L(3)\} = K\{L(2)\} + 1 = 3 + 1 = 4 \,,$$
$$K\{L(4)\} = K\{L(3)\} + 1 = 4 + 1 = 5 \,,$$

etc. One immediately sees that

$$K\{L(h)\} = h + 1 \,. \tag{5.6}$$

A similar procedure gives for the zig-zag chain

$$K\{A(3)\} = K\{A(2)\} + K\{A(1)\} = 3 + 2 = 5 \,,$$
$$K\{A(4)\} = K\{A(3)\} + K\{A(2)\} = 5 + 3 = 8 \,,$$

etc. One arrives at the sequence of numbers 2, 3, 5, 8, 13, 21, 34, 55, 89, ... , which in combinatorial mathematics are known under the name of FIBONACCI numbers,

F_k. These are defined as $F_0 = F_1 = 1$, $F_k = F_{k-1} + F_{k-2}$ for $k \geq 2$. In the general case,

$$K\{A(h)\} = F_{h+1} \cdot$$

As a curiosity we note that the FIBONACCI numbers satisfy the following peculiar algebraic expression

$$F_k = \frac{1}{\sqrt{5}} \left[\left(\frac{1 + \sqrt{5}}{2} \right)^{k+1} - \left(\frac{1 - \sqrt{5}}{2} \right)^{k+1} \right].$$

Therefore

$$K\{A(h)\} = \frac{1}{\sqrt{5}} \left[\left(\frac{1 + \sqrt{5}}{2} \right)^{h+2} - \left(\frac{1 - \sqrt{5}}{2} \right)^{h+2} \right]. \tag{5.7}$$

Equations (5.6) and (5.7) are the *solutions* of the recurrence relations (5.4) and and (5.5), respectively.

Recurrence relations for the number of KEKULÉ structures, as well as their solutions, are known for a variety of classes of benzenoid systems (CYVIN and GUTMAN 1988).

5.1.3 The Numeral-in-Hexagon Algorithm

In some cases it is possible to associate a numeral with each hexagon of a benzenoid system, such that their sum is equal to K. It is customary to inscribe the numeral in the hexagon.

The first such numeral-in-hexagon recipe was described by GORDON and DAVISON (1952). It enables an easy calculation of K of arbitrary unbranched cata-condensed benzenoids. We present here a suitably modified version (CYVIN and GUTMAN 1986b) of the original algorithm.

(i)   Start from an arbitrary side (say, left) of the chain. Write an external numeral one to the left of the first hexagon.
(ii) Enter unity in all hexagons until the first kink.
(iii) Right after each kink the entry to be written into the next hexagon is the sum of the numerals for the preceding linear segment.
(iv) The same numeral is then entered into all hexagons until the next kink, and so on.

The procedure suggests a building-up process of the benzenoid. The sum of the numerals in all hexagons (plus the external unity) gives the K value for any benzenoid during the build-up.

The example below may serve as an illustration.

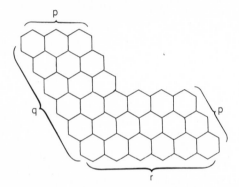

$$K = 4 \times 1 + 2 \times 3 + 4 \times 7 + 2 \times 31 = 100$$

Similar, though somewhat more complicated numeral-in-hexagon algorithms have been designed for a great number of other types of benzenoids (CYVIN and GUTMAN 1986b, 1988). To give the reader a flavour of these interesting methods, we present another relatively simple case, namely the ribbon $V(p, q, r)$:

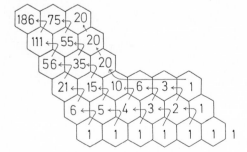

In our particular example it is $p = 3$, $q = 6$, $r = 6$. The algorithm starts with the insertion of unity into the hexagons of the lowest row and the most right (skew) column. The remaining hexagons are consecutively filled with numerals which are sums of previously determined numerals. The arrows in the below diagram indicate the summation rules.

By adding all the numerals thus obtained (together with the external unity) we obtain $K\{V(3, 6, 6)\} = 662$.

5.1.4 Combinatorial Formulas

Explicit combinatorial expressions for K are known for numerous classes of benzenoid systems (CYVIN and GUTMAN 1988). Two such formulas are (5.6) and (5.7). Here we provide some further examples of this kind (CYVIN and GUTMAN 1988).

Hexagon O(p, q, r)

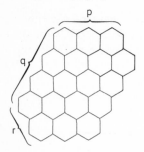

$$K\{O(p, q, r)\} = \prod_{i=0}^{q-1} \frac{\dbinom{p + r + i}{r}}{\dbinom{r + i}{r}} \tag{5.8}$$

The special case of (5.8) for $r = 1$ provides the K numbers of parallelograms.

Parallelogram L(p, q)

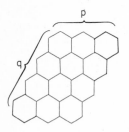

$$K\{L(p, q)\} = \dbinom{p + q}{q} \tag{5.9}$$

If, in addition, $q = 1$, then (5.9) reduces to (5.6).

Prolate rectangle $R^i(p, q)$

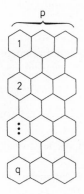

$$K\{R^i(p, q)\} = (p + 1)^q$$

Oblate rectangle $R^j(p, q)$

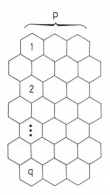

The problem of the enumeration of KEKULÉ structures of $R^j(p, q)$ has not yet been solved in terms of an explicit expression in p and q. Combinatorial formulas for $K\{R^j(p, q)\}$ are known (CYVIN and GUTMAN 1988) only for fixed values of $p, p \leq 8$, and for fixed values of $q, q \leq 4$. For instance

$$K\{R^j(4, q)\} = \frac{1}{20\,160} (q + 1) (q + 2)^4 (q + 3) \times$$

$$\times (17q^4 + 136q^3 + 439q^2 + 668q + 420)$$

$$K\{R^j(p, 4)\} = \frac{1}{27} \left[3^{p+1} + \frac{1}{4} (12 + 6 \sqrt{3})^{p+1} + \frac{1}{4} (12 - 6 \sqrt{3})^{p+1} \right].$$

5.1.5 The DEWAR-LONGUET-HIGGINS Formula

DEWAR and LONGUET-HIGGINS (1952) discovered an interesting relation between K and the adjacency matrix of a benzenoid system.

Consider a benzenoid system with n vertices and label them in an arbitrary manner with the numerals 1, 2, ... , n. The *adjacency matrix* A is a square matrix of order n whose (ij)-entry is unity if the vertices labeled by i and j are adjacent and is zero otherwise. Then (DEWAR and LONGUET-HIGGINS 1952),

$$\det A = (-1)^{n/2} K^2 . \tag{5.10}$$

This result holds for benzenoid systems only. The determinant of the adjacency matrix of other conjugated molecules may violate (5.10).

Equation (5.10) is particularly useful for the evaluation of K by means of computers, and suitable computer programs have been designed (BROWN 1983).

5.1.6 The JOHN-SACHS Formula

The notions of peak, valley, and monotonic path were defined in Chapter 3. Now, if a benzenoid systems has equal numbers of peaks and valleys, then a *monotonic path system* is a collection of independent (i.e. mutually nontouching) monotonic paths that involve all peaks and all valleys.

In Fig. 5.2 the monotonic path systems of benzo[a]pyrene are depicted.

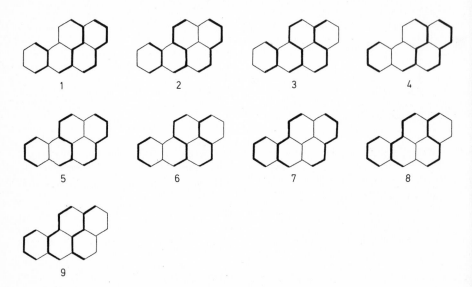

Fig. 5.2. The monotonic path systems of benzo[a]pyrene. The i-th monotonic path system corresponds to the i-th Kekulé structure in Fig. 5.1, $i = 1, 2, ... , 9$; for explanation, see text

GORDON and DAVISON (1952) observed that there is a one-to-one correspondence between a monotonic path system and a KEKULÉ structure. This is easy to see if one substitutes each monotonic path with a sequence of alternating double and single bonds, starting with a double bond. The remaining double bonds can then be positioned in a unique manner, as clarified by the example below.

monotonic path   Kekulé structure
   system

This argument leads to the following result, which was recently rigorously proved (SACHS 1984):

$$K = \# \text{ monotonic path systems} . \qquad (5.11)$$

Equation (5.11) automatically implies that benzenoids having an unequal number of peaks and valleys (thus having $\Delta \neq 0$) are non-Kekuléan.

JOHN and SACHS, and somewhat later JOHN and REMPEL, reported a further result of this kind.*

Consider benzenoid systems having an equal number r of peaks and valleys. Label (in arbitrary order) both the peaks and valleys with $1, 2, \ldots, r$. Let there be W_{ij} monotonic paths connecting the i-th peak with the j-th valley. Denote by W the matrix whose (ij)-entry is W_{ij}. Hence, W is a square matrix of order r.

The formula of JOHN and SACHS reads:

$$K = |\det W| . \qquad (5.12)$$

To make (5.12) more familiar, we present the monotonic paths connecting the peak 2 and the valley 1 of benzo[a]pyrene. Here and in Fig. 5.3 a less usual, but for our purpose more convenient, orientation of benzo[a]pyrene is used, with $r = 2$.

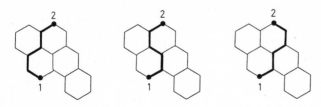

---

* The original papers by JOHN et al. were published in 1985 in two not easily available conference proceedings. Therefore the interested reader is referred to GUTMAN and CYVIN (1987) and CYVIN and GUTMAN (1988).

A simple method has been developed (Gutman and Cyvin 1987) by which the matrix elements W_{ij} are easily calculated. According to Gutman and Cyvin (1987), W_{ij} is equal to the number of Kekulé structures of the intersection graph I_{ij}. On the other hand, I_{ij} is the graph induced by those vertices that can be reached both from the i-th peak by going downwards and from the j-th valley by going upwards. Intersection graphs are usually benzenoids themselves and the determination of their K numbers is often an easy task.

In Fig. 5.3 are depicted the intersection graphs needed for the construction of the W-matrix of benzo[a]pyrene.

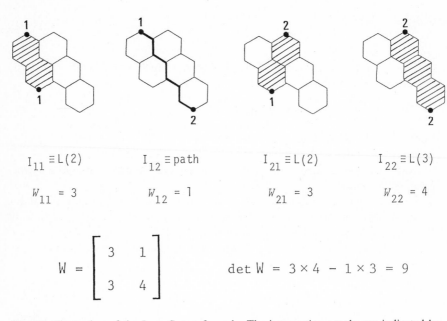

$$I_{11} \equiv L(2) \qquad I_{12} \equiv path \qquad I_{21} \equiv L(2) \qquad I_{22} \equiv L(3)$$

$$W_{11} = 3 \qquad W_{12} = 1 \qquad W_{21} = 3 \qquad W_{22} = 4$$

$$W = \begin{bmatrix} 3 & 1 \\ 3 & 4 \end{bmatrix} \qquad \det W = 3 \times 4 - 1 \times 3 = 9$$

Fig. 5.3. Illustration of the John-Sachs formula. The intersection graphs are indicated by hatching or (in the case of I_{12}) by a heavy line. The matrix elements W_{11}, W_{21} and W_{22} are just the K numbers of naphthalene, naphthalene and anthracene, respectively

5.2 The Case of $K = 0$

From Chapter 3 we know that the coloring of the vertices of a benzenoid system (with two colors) is achieved if all neighbors of a black vertex are white and vice versa. An immediate consequence of this is that all double bonds in a Kekulé structure lie between vertices of different colors. Therefore the existence of a

KEKULÉ structure implies that there are an equal number of black and white vertices, i.e. the color excess Δ is zero.

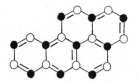 double bonds in a Kekulé structure connect differently colored vertices; therefore $K \neq 0 \Rightarrow \Delta = 0$

Benzenoid systems with non-zero color excess are necessarily non-Kekuléan. Because it is very easy to determine the color excess (e.g. by counting the peaks and valleys; see Section 3.4), these non-Kekuléans are termed "obvious". On the other hand, there exist non-Kekuléans with zero color excess. Their non-Kekuléan nature is much more difficult to reveal and therefore they are called "concealed".

In what follows we consider only benzenoid systems with an even number of vertices. The corresponding benzenoid hydrocarbons, provided they exist, possess an even number of π-electrons and therefore their total number of electrons is also even.

The opinion that non-Kekuléan conjugated π-electron systems should be polyradicalic species seems to be first stated explicitly in 1935 in an article by EUGEN MÜLLER and INGE MÜLLER-RODLOFF, but in a form that is not easily understandable. A clear formulation of the same principle is due to WHELAND (1944, 1955): *„A molecule is a diradical if, and only if, no structures of the conventional type can be written for it without formal bonds."* (Note that the work of MÜLLER and MÜLLER-RODLOFF (1935) is quoted by WHELAND (1955), but not by WHELAND (1944).) Anyway, the rule is nowadays known as MÜLLER-MÜLLER-RODLOFF's rule. In 1950 LONGUET-HIGGINS offered a simple rationalization of the rule within the framework of the molecular orbital theory. According to it, whenever the number of carbon atoms is even and $K = 0$, the highest occupied molecular orbitals are non-bonding and degenerate, each non-bonding molecular orbital being occupied by only one electron. The result is an open-shell (i.e. polyradicalic) π-electron configuration. The conclusions of LONGUET-HIGGINS (1950) apply to all alternant hydrocarbons and are thus valid for benzenoids.

In the earliest works on π-electron biradicals, benzenoid hydrocarbons were not considered. It was ERIC CLAR who first recognized even non-Kekuléan benzenoid hydrocarbons, the simplest representatives of which are triangulene, $C_{22}H_{12}$ (I), dibenzo[de,hi]naphthacene, $C_{24}H_{14}$ (II) and dibenzo[de,jk]pentacene, $C_{28}H_{16}$ (III). All these molecules have $\Delta = 2$ and therefore belong to the group of obvious non-Kekuléan benzenoids.

All attempts (CLAR and STEWART 1953) to synthesize I as well as the compounds II and III (CLAR, KEMP, and STEWART 1958) were unsuccessful, leading CLAR to the important conclusion that "KEKULÉ *structures are of paramount importance for the stability of aromatic ring systems*" (CLAR, KEMP, and STEWART 1958). Eventually, CLAR elaborated the concept of open-shell benzenoid systems (CLAR 1964a,

1964b, 1972). In CLAR (1964b) it was for the first time observed that the non-Kekuléan nature of I–III is stipulated by the fact that for these molecules the color excess is non-zero.

The existence of concealed non-Kekuléan benzenoid systems was first observed by CLAR (1972), who demonstrated this fact on example IV. In GUTMAN (1974) the systems IV and V were put forward. They both have eleven hexagons. It has been claimed by GUTMAN (1974) that eleven is the smallest possible value for h in the case of concealed non-Kekuléans. In a later work (BRUNVOLL, CYVIN SJ, et al. 1987) this was confirmed, and it was found that the eight benzenoids IV–XI are the only concealed non-Kekuléans with $h = 11$.

Concerning the chemistry of $K = 0$ benzenoid hydrocarbons, the situation is the following.

Not a single non-Kekuléan benzenoid hydrocarbon, either obvious or concealed, has ever been obtained.Not one has ever been detected as a short-lived transient (but existing!) species.

In short: non-Kekuléans do not exist (in 1988). However, this is not the whole truth and the reader should look at Section 9.2.

CLAR and STEWART (1953) made several attempts to synthesize triangulene, I. They first obtained hydroxytriangulenequinon, XII, a blue stable compound known since 1924. Reduction of XII by means of zinc dust fusion yielded hexahydrotriangulene, XIII. The idea was to dehydrogenate XIII and arrive at I.

This is what CLAR and STEWART (1953) report:

"When XIII was dehydrogenated by subliming it over a palladium-charcoal catalyst at 310° in a vacuum, it was absorbed completely into the catalyst and no sublimate was obtained. Even when the catalyst was heated to 500° no sublimate was obtained and it must be therefore assumed that polymerization was complete."

Eventually, an attempt was made to dehydrogenate dodecahydrotriangulene (XIV), employing a mild and cautious procedure – melting with palladium-charcoal at 200 °C under pure carbon dioxide; only XIII was obtained.

Further, CLAR and STEWART (1953) write:

"On heating to 250° a new evolution of hydrogen was observed, but at a greatly reduced rate. After 5 hours, high vacuum sublimation of the residue yielded only some XIII and no other triangulene derivatives could be detected."

Synthetic organic chemists have produced in recent years a great number of highly unusual compounds. It would be challenging for them to try their skills and the power of their new methods and new instruments on such a hard problem as the preparation of a non-Kekuléan benzenoid hydrocarbon. Some progress along these lines has already been made, see Section 9.2.

5.3 Resonance Theory

In the early period of quantum chemistry essentially two theories were developed, both being able to describe the electronic structure of conjugated, and, of course, other (both organic and inorganic), molecules. These two approaches are the valence bond (VB) and the molecular orbital (MO) theories. Although the basic notions of VB theory are much closer to the classical ideas of chemical structure, the contest was won by the MO theory. This occurred mainly because the mathematical and computational efforts required in a VB approach are significantly greater than those needed in an MO consideration of comparable accuracy.

The obvious obstacle of the VB theory is the enormous number of structures which must be taken into consideration. For instance, a complete VB description of benzene requires 5 covalent structures (two KEKULÉ-type and three DEWAR-type) and 170 ionic structures, 175 structures altogether. Since it is commonly accepted that the KEKULÉ-type VB structures are the most important ones, a radical resolution of the difficulty was achieved by neglecting everything except KEKULÉ structures. This crudest possible simplification of the VB theory is called resonance theory (RT). Surprisingly enough, it works.

Resonance theory flourished in the 1940s and 1950s, but after that it gradually lost its popularity among organic chemists. There are several reasons for this, one of which is certainly the fact that deliberations based on RT provided (at that time) only a qualitative description of (some) chemical facts. In the early 1970s RT was considered by the majority of scientists as an obsolete form of chemical reasoning, which perhaps could still be used in the pedagogy of organic chemistry.

In 1973 WILLIAM HERNDON put forward a simple *quantitative* RT approach, by which — using only KEKULÉ structures — one could predict a number of properties of benzenoid (and, more generally, of conjugated) hydrocarbons with accuracy tantamount to the best SCF MO calculations (SCF = Self-Consistent Field). Although the basic ideas of this approach can be found in the works of W. T. SIMPSON (see SIMPSON 1953, 1956 and the references cited therein), it is justified to speak about HERNDON's resonance theory (HRT).

In this section we briefly outline the basic features of HRT and point out its practical applicability in the theory of benzenoid hydrocarbons. The success of HRT will be documented by numerous examples. The reason for this success remains obscure. No satisfactory quantum mechanical (or any other) argument has been offered to explain how such an apparently rough and unjustified approximation can operate.*

---

* Recently KLEIN and SCHMALZ (1989) demonstrated that HERNDON's model can be rationalized by means of resonance-theoretical cluster expansion. On the other hand HERNDON's model was found to be not fully deductable from either the PAULING-WHELAND resonance theory or the bond-orbital resonance-theoretical approach of ŽIVKOVIĆ (KLEIN and TRINAJS-TIĆ 1989).

The basic assumption in HERNDON's resonance theory is that KEKULÉ structures alone suffice for the description of a conjugated system. To each KEKULÉ structure k_i a formal wave function $|k_i\rangle$ is associated*, $i = 1, 2, \ldots, K$. It is further assumed that the $|k_i\rangle$s are mutually orthogonal and that they contribute with equal weight to the (ground state) wave function $|\Psi\rangle$ of the molecule:

$$|\Psi\rangle = K^{-1/2} \sum_{i=1}^{K} |k_i\rangle \, .$$

This gives for the total energy of the π-electrons:

$$E_\pi = \frac{1}{K} \sum_{i=1}^{K} \langle k_i| \hat{H} |k_i\rangle + \frac{2}{K} \sum_{i<j} \langle k_i| \hat{H} |k_j\rangle \, ,$$

where $\hat{H}$ stands for some formal Hamiltonian operator (whose form is not specified and which again needs not have much in common with a VB Hamiltonian). It is reasonable to assume that $\langle k_i| \hat{H} |k_i\rangle$, the energy of the i-th KEKULÉ structure is independent of i. Then (HERNDON and ELLZEY 1974)

$$E_\pi = \langle k| \hat{H} |k\rangle + \frac{2}{K} \sum_{i<j} \langle k_i| \hat{H} |k_j\rangle \, , \tag{5.13}$$

and the second term on the right-hand side of (5.13) can be interpreted as the resonance energy (= the energy gain or energy loss due to the interaction between KEKULÉ structures).

The matrix elements $\langle k_i| \hat{H} |k_j\rangle$ determining the resonance energy obviously depend on the two KEKULÉ structures k_i and k_j. HERNDON proposed that their interaction depends only on the number of double bonds which have to be cyclically permuted in order to convert k_i into k_j or vice versa. If the number of cyclically permuted double bonds is $2\lambda + 1$, then

$$\gamma_\lambda = \langle k_i| \hat{H} |k_j\rangle = (\gamma_2/\gamma_1)^{\lambda-1} \gamma_1 \, ,$$

where γ_1 and γ_2 were determined semiempirically as $\gamma_1 = 0.841$ eV, $\gamma_2 = 0.336$ eV (see HERNDON 1980 and the references cited therein). In practical applications the parameters γ_3, γ_4, etc. are set equal to zero.

In benzenoid systems every cyclic permutation of double bonds in a KEKULÉ structure requires the participation of an odd number ($= 2\lambda + 1$) double bonds. This important fact is demonstrated in the next chapter.

All this leads finally to a simple formula:

$$RE = \frac{2}{K}(\chi_1\gamma_1 + \chi_2\gamma_2) \, , \tag{5.14}$$

---

* It is by no means necessary to identify $|k_i\rangle$ with a basis function of VB theory (ŽIVKOVIĆ 1986). Moreover, the properties required of $|k_i\rangle$ differ significantly from those of a VB basis function.

where χ_1 and χ_2 stand for the number of pairs of KEKULÉ structures which are transformed one into the other by cyclically permuting 3 and 5 double bonds, respectively. The values of γ_1 and γ_2 are given above.

The application of formula (5.14) is rather simple as illustrated in Fig. 5.4. There exist simple graph-theoretical techniques by which the numbers χ_1 and χ_2 can be determined easily; they are presented in some more detail in the next chapter (see Section 6.5).

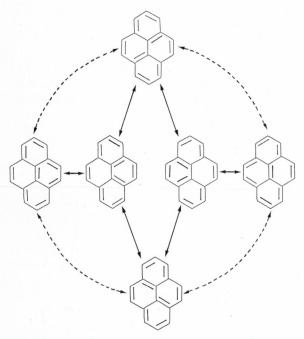

Fig. 5.4. HERNDON-type interactions between the KEKULÉ structures of pyrene. Full lines: interactions involving three double bonds; dashed lines: interactions involving five double bonds; higher order interactions are not indicated. Therefore $RE = (2/6)\,(6 \times \gamma_1 + 4 \times \gamma_2) =$ $= 2.130\ \mathrm{eV}$

It is remarkable that formula (5.14), in spite of its extreme simplicity and obvious resonance-theoretical heritage, excellently reproduces the resonance energies computed by an advanced (and highly parametrized) SCF MO method (DEWAR and DE LLANO 1969), which are commonly accepted as the best resonance energies available*. A few selected examples are collected in Table 5.1.

---

* Discussions about resonance energy and the closely related concept of aromaticity go beyond the ambit of the present book. Readers interested in this controversial matter should consult BADGER 1969, BINSCH 1973, GEORGE 1975 and TRINAJSTIĆ 1983, Vol. II, pp. 1—4.

Table 5.1. Resonance energies of some benzenoid hydrocarbons; all values are in eV

Compound	DEWAR & DE LLANO (1969)	Eq. (5.14)	Eq. (5.15)
Benzene	0.87	0.84	0.82
Naphthalene	1.32	1.35	1.30
Anthracene	1.60	1.60	1.64
Phenanthrene	1.93	1.95	1.91
Pyrene	2.10	2.13	2.12
Chrysene	2.48	2.52	2.46
Triphenylene	2.65	2.65	2.60
Perylene	2.62	2.69	2.60
Pentacene	2.00	1.85	2.12
Benzo[a]pyrene	2.58	2.58	2.60
Benzo[e]pyrene	2.85	2.87	2.84
Coronene	3.52	3.50	3.55

As a further surprise, it was later shown (SWINBORNE-SHELDRAKE et al. 1975) that a similar accuracy can be achieved by an even simpler formula, viz.

$$RE = 1.185 \ln K \text{ [eV]} \tag{5.15}$$

Some illustrative examples can be found in Table 5.1.

Eventually, a large number of applications of HRT was reported.

(i) The standard (gas phase) heats of atomizations of benzenoid hydrocarbons can be calculated as (HERNDON 1974b):

$$\Delta H_a^0(g) = 445.84 \times (\# \text{ CH bonds}) + 463.70 \times (\# \text{ CC bonds})$$
$$+ RE - 10.24 \times (\# \text{ bays}) - 44.89 \times (\# \text{ coves}) \text{ [kJ mol}^{-1}], \tag{5.16}$$

where RE is given by (5.14) and $\gamma_1 = 51.53$, $\gamma_2 = 19.07$ have been used. Note that in the notation introduced in Chapter 3, $\#$ CH bonds $= \#$ vertices of degree two $= 2h - 2$, $\#$ CC bonds $= \#$ edges $= m$. Equation (5.16) is equivalent to

$$\Delta H_f^0(g) = 23.49n - 7.99h + 463.70 - RE + 10.24 \times (\# \text{ bays})$$
$$+ 44.89 \times (\# \text{ coves}) \text{ [kJ mol}^{-1}] \tag{5.17}$$

for the standard heats of formation.

(ii) Adiabatic ionization potentials of benzenoid hydrocarbons are reproduced by means of (HERNDON 1976)

$$IP = 11.277 + 1.185 \ln K - 1.044 \ln K^* \text{ [eV]},$$

where K^* is the count of resonance forms of the respective radical cation (HERNDON 1975). Note that the validity of this approach was later strongly criticized (EILFELD and SCHMIDT 1981).

(iii) The SCF MO π-electron and $(\sigma + \pi)$-electron stabilization energy of aromatic radicals can be estimated by similar formulas (STEIN and GOLDEN 1977):

$$\pi RSE = 94.89 \ln K^* - 114.35 \ln K \, [\text{kJ mol}^{-1}] \,,$$
$$(\sigma + \pi) \, RSE = 70.08 \ln K^* - 84.47 \ln K \, [\text{kJ mol}^{-1}] \,.$$

(iv) Several types of chemical reactivity of benzenoid hydrocarbons were quantitatively described by HRT: electrophilic and nucleophilic reactions, solvolysis, rates of hydrogen isotope exchange, etc. Also the carcinogenic activity of benzenoid hydrocarbons is rationalized by means of HRT. For details on these matters, see HERNDON (1980).

A typical result of this kind is the formula (BIERMANN and SCHMIDT 1980)

$$\log k = 3.330 + 6.972 \ln (K_P/K_R)$$

for the rate constant (k) of the DIELS-ALDER reaction of benzenoid hydrocarbons with maleic anhydride. Here, K_P and K_R stand for the number of KEKULÉ structures of the product and the reactant, respectively. Although this formula gives results quite close to the experimental values, it is somewhat inferior to a similar expression derived from simple molecular orbital theory. (HESS et al. 1981).

(v) The prediction of bond lengths via resonance theory is discussed in a subsequent section.

5.4 KEKULÉ Structures and the Thermodynamic Stability of Benzenoid Hydrocarbons

We have already mentioned the work of HERNDON (1974b) where he expressed the heats of formation of benzenoid hydrocarbons as a linear function of the resonance energy and some other structural details; see (5.17). Comparing (5.17) and (5.15) one may conclude that the thermodynamic stability of benzenoid systems is proportional to the number of KEKULÉ structures. However, the situation with the dependence of the heats of formation on KEKULÉ structures is much more complex.

First of all, one should note that the heats of formation of benzenoid hydrocarbons can be expressed by means of group additivity terms, none of which has any connection with KEKULÉ structures (for details see Section 9.1). These group additivity schemes yield a slightly more accurate prediction of the enthalpies of benzenoid hydrocarbons than (5.16) and (5.17) (STEIN, GOLDEN, and BENSON 1977). Therefore the use of these latter equations is rather a matter of choice than a necessity which follows from theory.

Another way in which KEKULÉ structures can be connected with the thermodynamic stability of benzenoid hydrocarbons is via their molecular orbital total π-electron energies. For this purpose, the total π-electron energy calculated by the simple tight-binding approach of HÜCKEL is particularly convenient because our understanding of the dependence of this quantity on molecular structure is fairly good (see Chapter 12 in GUTMAN and POLANSKY 1986). In what follows, the HÜCKEL molecular orbital total π-electron energy, expressed as usual in units of β (GUTMAN and POLANSKY 1986), will be denoted by E.

It has been claimed (SCHAAD and HESS 1972) that there is a linear correlation between E and the heats of formation of benzenoid hydrocarbons. Indeed, such a correlation exists and its quality can be judged from Fig. 5.5.

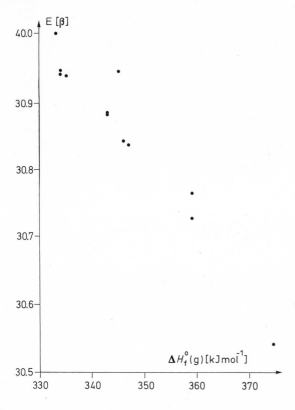

Fig. 5.5. Correlation between HÜCKEL molecular orbital total π-electron energy (E) and the standard gas phase heats of formation for the 12 isomeric catacondensed benzenoid hydrocarbons $C_{22}H_{14}$; the correlation coefficient is -0.966

Based upon the assumption that the basic structural factors determining the thermodynamic stability of benzenoid (and, more generally, conjugated) compounds are reflected in their E values, much theoretical work has been done to elucidate the dependence of E on molecular structure. Here we outline only the research concerned with KEKULÉ structures.

It seems that HALL (1973) was the first to observe a simple, but long overlooked fact, namely that within families of isomeric benzenoids an excellent linear correlation exists between E and K. Figure 5.6 provides a clear illustration of HALL's rule.

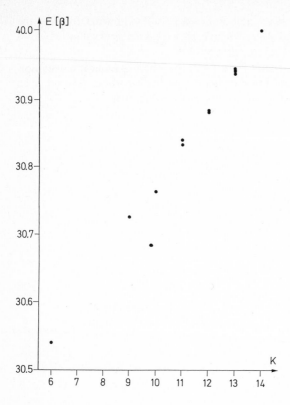

Fig. 5.6. Correlation between Hückel molecular orbital total π-electron energy (E) and the number of Kekulé structures for 12 isomeric catacondensed benzenoid hydrocarbons $C_{22}H_{14}$; the correlation coefficient is 0.9987

A theoretical explanation of Hall's rule was offered for catacondensed benzenoids (Gutman and Petrović 1983). For pericondensed benzenoids the reason for such a simple dependence of E on K is still obscure.

Recently Cioslowski (1986) and Cioslowski and Polansky (1988) put forward a theory according to which the total π-electron energy of benzenoid hydrocarbons must have the following functional dependence on n (number of carbon atoms), m (number of carbon-carbon bonds), and K:

$$E = \sqrt{2mn}\, F(x) \,,$$

where

$$x = K^{2/n}(2m/n)^{-1/2} \,.$$

Within Cioslowski's theory, the form of the function $F(x)$ remains unspecified, but it seems (Gutman, Marković and Marinković 1987) that $F(x)$ depends on x in a nearly linear manner. Assuming that $F(x) = ax + b$, the following approximate formula has been put forward (CIOSLOWSKI 1986, GUTMAN, MARKOVIĆ, and MARINKOVIĆ 1987):

$$E = 0.7578\sqrt{2mn} + 0.1899nK^{m/2} .\tag{5.18}$$

Equation (5.18) represents one of the most reliable approximations for E of benzenoid hydrocarbons and, among other things, reveals that the structural parameters which determine the gross part (some 99.9%) of E are n, m, and K.

The peculiar thing with (5.18) and with CIOSLOWSKI's theory in general is that it predicts a nonlinear dependence of E on K. Attempts have been made (GUTMAN 1986) to reconcile HALL's rule and CIOSLOWSKI's theory, but a satisfactory resolution of this controversy is not known.

5.5 KEKULÉ Structures and Bond Lengths

In each KEKULÉ structure the bonds between adjacent carbon atoms are either double or single. Since the real bonding in a conjugated molecule is much better represented by a multitude of KEKULÉ structures, it is natural to anticipate that the order of a π-bond between two carbon atoms lies between zero ($=$ pure single bond) and one ($=$ pure double bond). The most straightforward way to evaluate such a bond order is to take a simple, nonweighted, average over all KEKULÉ structures. This idea was first proposed by PAULING et al. (1935) and was eventually elaborated in detail (PAULING 1960, pp. 234—239).

In the case of naphthalene, the bond 1—2 is double in two and single in one KEKULÉ structure. Therefore a bond order 2/3 is associated with it. Similarly the bonds 2—3 and 5—10 are both of order 1/3. The bond order determined according to this method is called the PAULING bond order. In the general case, the PAULING bond order between the atoms r and s is defined as:

$$P_{rs} = K_{rs}/K ,$$

where K stands for the number of KEKULÉ structures and

$K_{rs} = \#$ KEKULÉ structures in which the atoms r and s are connected by a double bond.

The PAULING bond orders of benzo[a]pyrene are calculated as follows. The reader may easily obtain the numbers below by means of Fig. 5.1.

$P_{1\,2} = 5/9$ $\quad P_{9\,10} = 3/9$ $\quad P_{17\,18} = 2/9$

$P_{2\,3} = 4/9$ $\quad P_{10\,11} = 6/9$ $\quad P_{18\,19} = 3/9$

$P_{3\,4} = 5/9$ $\quad P_{11\,12} = 3/9$ $\quad P_{19\,20} = 3/9$

$P_{4\,5} = 1/9$ $\quad P_{12\,13} = 6/9$ $\quad P_{1\,18} = 4/9$

$P_{5\,6} = 8/9$ $\quad P_{13\,14} = 3/9$ $\quad P_{4\,19} = 3/9$

$P_{6\,7} = 1/9$ $\quad P_{14\,15} = 3/9$ $\quad P_{7\,20} = 2/9$

$P_{7\,8} = 6/9$ $\quad P_{15\,16} = 2/9$ $\quad P_{9\,14} = 3/9$

$P_{8\,9} = 3/9$ $\quad P_{16\,17} = 7/9$ $\quad P_{15\,20} = 4/9$

It is instructive to check the results by adding the orders of all bonds which end at an atom; the sum must always be equal to unity.

In addition to the PAULING bond order a variety of other bond-order-type quantities have been considered in the chemical literature. It should not be surprising that numerous relations exist between these bond orders. These relations have been systematically examined in GUTMAN (1977a), where additional information on bond orders can also be found. One of these relations deserves to be mentioned here, viz.

$$(A^{-1})_{rs} = P_{rs}, \tag{5.19}$$

where A is the adjacency matrix defined in Paragraph 5.1.5 in connection with (5.10). Equation (5.19) holds only for adjacent atoms of benzenoid hydrocarbons. It was discovered by HEILBRONNER (1962) and, in a somewhat different form, by HAM (1958). An application of (5.19) can be found in Section 4.5.

The definition of any bond-order-like quantity is somewhat arbitrary because bond orders are not experimentally measurable. However, the closely related quantities — the equilibrium distances between the nuclei of the bonded atoms — are obtainable from X-ray and neutron diffraction experiments for the crystalline phase, as well as from gas electron diffraction, and are indeed known for quite a few benzenoid hydrocarbons.

It has been shown many times that the simple PAULING bond order can be used to predict bond lengths with an accuracy which is comparable to that of the available experimental data (PAULING 1960, HERNDON 1974a, HERNDON, and PÁRKÁ-NYI 1976, PAULING 1980). PAULING (1960, 1980) proposed the formula

$$D_{rs} = D_0 - (D_0 - D_1) \frac{1.84 P_{rs}}{0.84 P_{rs} + 1} \text{ [pm]},$$ (5.20)

in which D_{rs} is the distance between the atoms r and s, D_0 is the length of a single bond between sp^2 hybridized carbon atoms ($= 150.4$ pm), and D_1 is the length of a double bond ($= 133.4$ pm).

In Table 5.2 we present the experimental and calculated bond lengths in tetra-benzoheptacene. It can be seen that (5.20) is accurate to about 1 pm, which is close to the experimental precision. In the example shown in Table 5.2 the mean deviation between experimental and theoretical bond lengths is 1.2 pm. Such a deviation for bond lengths obtained using molecular orbital (COULSON-type) bond orders is two times greater (PAULING 1980).

Table 5.2. Bond lengths in tetrabenzo[de, no, st, c_1, d_1]heptacene. The calculated bond lengths are obtained from the PAULING bond order using (5.20). All bond lengths are given in pm; all data are taken from PAULING (1980)

Bond	PAULING bond order	Bond length calc.	Bond length exp.
a	0.809	135.4	134
b	0.191	145.2	144
c	0.427	140.7	142
d	0.573	138.4	136
e	0.427	140.7	140
f	0.573	138.4	139
g	0.382	141.3	142
h	0.382	141.3	141
i	0.236	144.2	143
j	0.045	149.0	148
k	0.227	144.4	146
l	0.727	136.4	136
m	0.273	143.6	142
n	0.227	144.4	142
o	0.500	139.4	138

HERNDON (1974a) and HERNDON and PÁRKÁNYI (1976) preferred to apply a linear relation between P_{rs} and D_{rs}. Their formula

$$D_{rs} = 146.4 - 12.4P_{rs} \text{ [pm]}$$

was obtained by adjusting the numerical coefficients to only four π-electron systems: ethylene, butadiene, benzene and graphite. It reproduces the experimental data with a similar accuracy as (5.20) and is shown to be superior to analogous formulas based on molecular orbital bond orders.

Hence, by counting the KEKULÉ structures and using the PAULING bond order concept it is possible to predict the geometry of benzenoid hydrocarbons (even of very large ones) with a precision which is usually not worse than that of other theoretical approaches. The great advantage of using PAULING bond orders is the ease with which the computations can be performed — they require only a few minutes of paper-and-pencil work. Efficient graph-theoretical algorithms for calculation of P_{rs} have been developed (HERNDON 1974a, RANDIĆ 1975a, HERNDON and PÁRKÁNYI 1976).

At the end of this section we mention the correlation which exists between NMR spin-spin coupling constants of *ortho*-hydrogen atoms in benzenoid hydrocarbons and the respective PAULING bond orders. It has been demonstrated (HERNDON 1974a) that the formula

$$J_{HH}^{ortho} = 5.27 + 4.34P_{rs} \text{ [Hz]} \tag{5.21}$$

reproduces the coupling constants of hydrogen atoms attached to (adjacent) atoms r and s with an error of about 0.1 Hz. If one of the atoms r, s lies in a bay, then a correction term of 0.54 Hz is to be added to the right-hand side of (5.21).

5.6 KEKULÉ Structures and other Physico-Chemical Properties of Benzenoid Hydrocarbons

A plethora of reports exists in the chemical literature about correlations of various physico-chemical properties of benzenoid hydrocarbons with the number of KE-KULÉ strctures. Some of these works have already been mentioned in the preceding parts of this chapter; some more are revealed in the following.

Generally speaking, approaching the results collected in this section requires a certain amount of caution. They are often obtained from the analysis of a limited data base. Critical comparison with other rival approaches may be missing. From a statistical point of view, some of the correlations are not at all convincing.

JOELA (1975) associated to each KEKULÉ structure of a benzenoid hydrocarbon a graph whose vertices represent the double bonds. This graph is sometimes called a *factor graph*. Its construction is evident from the example below. The eigenvalues of the factor graphs (of all KEKULÉ structures) were correlated with the wavenumbers of the absorption spectrum of the respective benzenoid radical anion. A linear correlation was established.

Kekulé structure   Joela's factor graph

In searching for an index which would characterize the local aromaticity of a benzenoid hydrocarbon, RANDIĆ (1975b) introduced the so-called *local ring index*

$$R(B, H) = 2K\{B - H\}/K\{B\} .$$

Here, H is a hexagon of the benzenoid system B and $B - H$ is the system obtained by deleting from B the six vertices of H. It was argued that the methyl doublet separations in the NMR spectra of *ortho*-dimethyl substituted benzenoid hydrocarbons are correlated with the respective local ring index. The curvilinear relation reported by RANDIĆ (1975b) is based on five experimental values.

The sum of the local ring indices over all hexagons is interpreted as an *overall aromaticity index* of the respective benzenoid hydrocarbon (RANDIĆ 1975b). BRÄUCHLE et al. (1980) correlated this overall index of RANDIĆ with the triplet zero-field splitting parameters of linear polyacenes and pyrene derivatives. The results were of comparable quality to those with other molecular orbital indices. Separate correlations were made for polyacenes and for pyrenes. For samples containing benzenoid hydrocarbons of various types, the correlations were quite poor, and therefore a much more complicated model was eventually proposed (BRÄUCHLE and VOITLÄNDER 1982). The model put forward by BRÄUCHLE and VOITLÄNDER (1982) is also based on KEKULÉ structure enumeration and requires the K values of certain fragments of the benzenoid hydrocarbon considered.

Chapter 6

Conjugated Circuits

6.1 Alternating Cycles

In Chapter 3 we saw that the mathematical object corresponding to a KEKULÉ structure is the 1-factor. The theory of 1-factors is well elaborated (LOVÁSZ and PLUMMER 1986), and in this field numerous mathematical results have been established. In this section we quote some of them, in particular those which provide the basis of the conjugated circuit model.

As already defined in Chapter 3, a 1-factor of a graph G is a collection of edges of G such that every vertex of G is covered by exactly one edge belonging to the 1-factor. Let k_i and k_j be two distinct 1-factors of a graph G. Consider the collection $k_i \Delta k_j$ of those edges which belong either to k_i or to k_j, but not to both k_i and k_j. (In mathematics $k_i \Delta k_j$ is called the *symmetric difference* of k_i and k_j.)

It is easy to see that $k_i \Delta k_j$ is composed of disjoint cycles (or is simply a cycle). Furthermore, all cycles in $k_i \Delta k_j$ are of even size. Below are given two examples in which $k_i \Delta k_j$ contains 3 and 1 cycle, respectively.

$$k_1 \qquad k_2 \qquad k_3 \qquad k_1 \Delta k_2$$

$$k_1 \Delta k_3$$

Hence, $k_1 \Delta k_2$ consists of one ten-membered and two six-membered cycles; $k_1 \Delta k_3$ consists of a single 18-membered cycle. At this point it would be useful for the reader to construct $k_2 \Delta k_3$ and to verify that it also possesses just one cycle.

A cycle of $k_i \, \Delta k_j$ is said to be an *alternating cycle* of k_i and of k_j. This name is appropriate: in a 1-factor the edges belonging to an alternating cycle are alternately double and single. (For the terms "double edge" and "single edge" of a 1-factor, see Chapter 3.)

If a 1-factor is given, then it is not easy to recognize its alternating cycles. The following procedure may then be applied.

Let $k_1, k_2, \ldots, k_K$ be the 1-factors of a graph. Construct $k_i \, \Delta k_1, \, k_i \, \Delta k_2, \ldots,$ $k_i \, \Delta k_K$. The cycles which occur in any of these symmetric differences (and only these cycles) are the alternating cycles of k_i.

The following result is also often used in the conjugated circuit model. It holds for graphs which possess 1-factors.

Let G be a graph and Z its cycle. Denote by $G–Z$ the subgraph obtained by deletion of the vertices of Z from G. Then Z is an alternating cycle of (at least two) 1-factors of G if and only if $G–Z$ possesses 1-factors. (In the special case when Z is the Hamiltonian cycle of G and when $G–Z$ possesses no vertices, Z is an alternating cycle.)

Alternating cycles and symmetric differences of 1-factors/KEKULÉ structures have been investigated in theoretical chemistry on many occasions (see GRAOVAC et al. 1972, CVETKOVIĆ et al. 1974, GUTMAN and HERNDON 1975 and the references cited therein). However, it was MILAN RANDIĆ (1976) who first recognized the great significance of alternating cycles for the chemical behaviour of conjugated molecules. He named them "*conjugated circuits*" and developed a theory which is known under the name "*the conjugated circuit model*".

6.2 The Conjugated Circuit Model

The theory outlined in this chapter has to be associated with the name of MILAN RANDIĆ who discovered it (RANDIĆ 1976) and eventually elaborated it (RANDIĆ 1977a, 1977b) and applied it to numerous classes of conjugated molecules (RANDIĆ 1980, 1982, RANDIĆ, SOLOMON, et al. 1987, and the references cited therein).

In what follows we describe only the conjugated circuit model for benzenoid hydrocarbons. One should note, however, that the model covers a much wider class of conjugated systems (RANDIĆ 1977a, 1977b, 1982).

The basic assumption of the conjugated circuit model is that the conjugated circuits in the KEKULÉ structures (= the alternating cycles in the 1-factors) are the dominant features determining the stabilities and, in particular, the resonance energies of conjugated molecules. It has been shown (RANDIĆ 1976) that the reso-nance energies are simple additive functions of the conjugated circuits, and only the size of a circuit has to be taken into account.

First of all one has to recognize and count the conjugated circuits in the KE-KULÉ structures. This is called (RANDIĆ 1976, 1977b) *circuit decomposition*. We illustrate this concept with the example of benzo[a]pyrene, whose nine KEKULÉ structures can be found in Fig. 5.1.

Examine first the KEKULÉ structure 1:

1

R$_1$ R$_3$ R$_4$ R$_2$

R$_1$ R$_4$

A trial-and-error procedure (which is by no means simple, and requires some practice) reveals that in this KEKULÉ structure there are two 6-membered, one 10-membered, one 14-membered and two 18-membered conjugated circuits.

The case of the KEKULÉ structure 2 is somewhat simpler:

2

R$_1$ R$_1$ R$_1$ R$_2$

R$_4$

In the theory put forward by RANDIĆ some conjugated circuits are considered linearly dependent. For example, one of the 18-membered conjugated circuits of the KEKULÉ structure 1 of benzo[a]pyrene can be obtained by adding and subtracting some other conjugated circuits, which are also present in 1:

If a conjugated circuit can be expressed as a linear combination of conjugated circuits of smaller size, then it is disregarded in RANDIĆ's theory. It is, of course, assumed that the conjugated circuits among which a linear relation exists are all contained in one and the same KEKULÉ structure.

Those conjugated circuits which cannot be expressed as linear combinations of conjugated circuits of smaller size (of the same KEKULÉ structure) are considered *linearly independent*. In particular, 6- and 10-membered conjugated circuits are always independent.

An inspection of the KEKULÉ structures 1 and 2 reveals that the above mentioned linear relation is the only one. Hence, the KEKULÉ structure 1 contains two 6-membered, one 10-membered, one 14-membered and one 18-membered independent conjugated circuits. This will be written symbolically as

$$k_1 = 2R_1 \oplus R_2 \oplus R_3 \oplus R_4 ,$$

where k_1 stands for the KEKULÉ structure 1 and R_λ is the conjugated circuit of size $4\lambda + 2$.

Similarly,

$$k_2 = 3R_1 \oplus R_2 \oplus R_4 ,$$

in harmony with the fact that all the five conjugated circuits of 2 are independent.

Repeating the same analysis for the remaining KEKULÉ structures of benzo[a]-pyrene (see Fig. 5.1) we obtain:

$$k_3 = 2R_1 \oplus 2R_2 \oplus R_3 ,$$
$$k_4 = 2R_1 \oplus 2R_2 \oplus R_3 ,$$
$$k_5 = 4R_1 \oplus R_3 ,$$
$$k_6 = 3R_1 \oplus 2R_2 ,$$
$$k_7 = 2R_1 \oplus 3R_2 ,$$
$$k_8 = 3R_1 \oplus R_2 \oplus R_3 ,$$
$$k_9 = R_1 \oplus 2R_2 \oplus 2R_3 .$$

Summing up these nine equations we obtain the *circuit decomposition* of benzo[a]-pyrene (into linearly independent conjugated circuits):

$$22R_1 \oplus 14R_2 \oplus 7R_3 \oplus 2R_4$$

or, as it is customary to write,

$$(22R_1 \oplus 14R_2 \oplus 7R_3 \oplus 2R_4)/9 ,$$

where "/9" stands for the nine KEKULÉ structures.

At this point we note the following two features. Firstly, the circuit decomposition of a benzenoid hydrocarbon is a quite cumbersome and error prone task. All KEKULÉ structures are to be investigated one-by-one, which for larger benzenoids becomes completely infeasible. The calculation of the numbers of R_1 and R_2 circuits (which necessarily are independent) is somewhat facilitated when suitable graph-theoretical algorithms are employed, as shown in Section 6.5. The analogous algorithms for finding the numbers of R_λ, $\lambda > 2$ (where linear dependence may be encountered) are significantly more complicated.

Secondly, in the study of benzo[a]pyrene we have found only conjugated circuits of size $4\lambda + 2$, $\lambda = 1, 2, 3, 4$. This, as we prove in the next section, is a generally valid property of benzenoid hydrocarbons.

It is particularly difficult to recognize the linear dependency of the conjugated circuits, especially because of the lack of a general and systematic recipe for doing this. In addition to the example of R_4 considered above, in benzo[a]pyrene, we point out the linear dependency of the circuit R_5 in the KEKULÉ structure of benzo[a]-coronene below:

It reads:

Finding the above relation would be a difficult task even for an expert.

The circuit decomposition of an arbitrary benzenoid system B (which, of course, must be Kekuléan) will be written as

$$(\varrho_1 R_1 \oplus \varrho_2 R_2 \oplus \cdots \oplus \varrho_A R_A)/K\{B\} , \tag{6.1}$$

where $4\Lambda + 2$ is the size of the largest independent conjugated circuit of B.

In the case of benzo[a]pyrene, $\Lambda = 4$ and $\varrho_1 = 22$, $\varrho_2 = 14$, $\varrho_3 = 7$, $\varrho_4 = 2$.

It is instructive to derive the circuit decomposition of benzene, naphthalene, anthracene and naphthacene, which can be recommended as a relatively easy exercise for the reader.

$(2R_1)/2$ $(4R_1 \oplus 2R_2)/3$ $(6R_1 \oplus 4R_2 \oplus 2R_3)/4$

$(8R_1 \oplus 6R_2 \oplus 4R_3 \oplus 2R_4)/5$

RANDIĆ (1976) has used the circuit decomposition (6.1) to calculate the resonance energy of B according to the formula

$$RE = (\varrho_1 R_1 + \varrho_2 R_2 + \cdots + \varrho_A R_A)/K\{B\} , \tag{6.2}$$

where R_λ, $\lambda = 1, 2, \ldots, \Lambda$ is an additive increment, representing the contribution of a $(4\lambda + 2)$-membered conjugated circuit to the resonance energy of the respective molecule.

The parameters R_λ were determined in the following manner. R_1, R_2, R_3 and R_4 were calculated so as to exactly reproduce the SCF MO resonance energies of DEWAR and DE LLANO (1969) of benzene, naphthalene, anthracene and naphthacene. In other words, R_1, R_2, R_3, and R_4 were the solutions of the following system of equations:

$(2R_1)/2 = RE \text{ (benzene)} = 0.869 \text{ eV} ,$
$(4R_1 + 2R_2)/3 = RE \text{ (naphthalene)} = 1.323 \text{ eV} ,$
$(6R_1 + 4R_2 + 2R_3)/4 = RE \text{ (anthracene)} = 1.600 \text{ eV} ,$
$(8R_1 + 6R_2 + 4R_3 + 2R_4)/5 = RE \text{ (naphthacene)} = 1.822 \text{ eV.}$

This gives:

$$R_1 = 0.869 \text{ eV} ,$$
$$R_2 = 0.246 \text{ eV} ,$$
$$R_3 = 0.100 \text{ eV} ,$$
$$R_4 = 0.140 \text{ eV} .$$

However, for an obscure reason (which, we are afraid, was simply a computational error)*, RANDIĆ (1976) accepted the value 0.041 eV for the parameter R_4. In all the numerous publications on the conjugated circuit model, including the book of TRINAJSTIĆ (1983), Vol. II, pp. 70–103, the same value for R_4, i.e.

$$R_4 = 0.041 \text{ eV}$$

has been employed. Furthermore, from this latter choice of R_4 it was concluded that the parameters R_λ rapidly decrease with increasing λ, and it was taken for granted that it is justified to set $R_5 = R_6 = \ldots = 0$.

Recently a new set of parameters ($R_1 = 0.869$ eV, $R_2 = 0.247$ eV, $R_3 = 0.100$ eV, and $R_\lambda = 0$, for $\lambda > 3$) has been proposed (RANDIĆ, NIKOLIĆ, and TRINAJSTIĆ 1987).

Irrespective of this pitfall, formula (6.2) was shown to reproduce the DEWAR – DE LLANO resonance energies of benzenoid hydrocarbons within a few hundredth of eV. Hence, (6.2) provides quite a satisfactory numerical value for the resonance energy.

For example, the resonance energy of benzo[a]pyrene is calculated as

$$RE = (22 \times 0.869 + 14 \times 0.246 + 7 \times 0.100 + 2 \times 0.041)/9 = 2.594 \text{ eV} ,$$

which is reasonably close to 2.584 eV as calculated by the SCF MO method.

6.3 The Basic Theorem

When the conjugated circuit model is applied to benzenoid hydrocarbons it is crucial that only $(4\lambda + 2)$-membered conjugated circuits occur. [In non-benzenoid conjugated molecules also (4λ)-membered conjugated circuits exist; their contribution to resonance energy is shown to be negative (RANDIĆ 1977a, 1977b).]

The $(4\lambda + 2)$-nature of the conjugated circuits in benzenoid systems was certainly known from the very beginning of the work on the conjugated circuit model (RANDIĆ 1976). Yet a general mathematical demonstration of this fact was never offered. (See, however, the results of CVETKOVIĆ et al. 1974.)

In this section we fill this gap.

We first demonstrate the validity of the equation

$$n = 4h + 2 - n_i, \quad \text{i.e.} \quad n + n_i = 4h + 2 \tag{6.3}$$

---

* Observe that 0.041 is "obtained" from 0.140 by exchanging a pair of digits.

connecting the numbers of vertices (n), hexagons (h) and internal vertices (n_i) of a benzenoid system. First of all, for $h = 1$ (i.e. for benzene, $n = 6$, $n_i = 0$) the relation is evidently true. Therefore it is sufficient to prove that if h increases by one, then $n + n_i$ increases by four.

In Chapter 2 we explained that a new hexagon can be added to a benzenoid system (which clearly increases the value of h by one) in five distinct ways. Consider each of these addition modes separately. The reader should look at the illustrations given in Chapter 2.

(i) In the case of a one-contact addition, four new vertices are added, but no new internal vertex is created. Therefore n increases by four and n_i remains the same. Hence, $n + n_i$ increases by four.

(ii) In the case of a two-contact addition, three new vertices are added, while one vertex which previously lay on the perimeter has become internal. Therefore n increases by three, and n_i increases by one. Hence, $n + n_i$ increases by four.

We leave it to the reader to verify that $n + n_i$ increases by four also in the case of three-, four- and five-contact additions. This will prove (6.3).

Next we show that if the perimeter of a benzenoid system has size 4λ (where λ is an integer), then n_i is odd. Furthermore, if the size of the perimeter is $4\lambda + 2$, then n_i is even.

Suppose that the perimeter consists of 4λ vertices. Then clearly $n = 4\lambda + n_i$. Using (6.3) we get $4\lambda + n_i = 4h + 2 - n_i$, i.e. $n_i = 2(h - \lambda) + 1$, and n_i is obviously an odd number.

If the perimeter has $4\lambda + 2$ vertices, then an analogous argument leads to $n_i = 2(h - \lambda)$, which means that n_i is an even number. This case also includes $n_i = 0$.

Note that the perimeter of any benzenoid system is either of size 4λ or of size $4\lambda + 2$.

An immediate corollary of the above result is that in the interior of any cycle of a benzenoid system there is an odd number of vertices if the size of the cycle is 4λ, and an even (or zero) number of vertices if the size of the cycle is $4\lambda + 2$. This conclusion is obtained by observing that any cycle in a benzenoid system can be understood as the perimeter of some smaller benzenoid system.

We are now prepared to deduce the "basic theorem" of the conjugated circuit model.

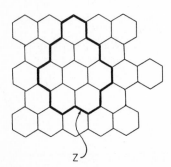

Z

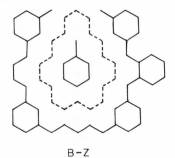

B-Z

Let Z be a cycle in the benzenoid system B. Since benzenoid systems do not contain odd-membered cycles, the size of Z is either 4λ or $4\lambda + 2$. By deleting Z from B it decomposes into two parts. One part contains the vertices from the interior of Z, the other the vertices from the exterior of Z. (These parts may be empty.)

Now, Z will be a conjugated circuit if and only if $B - Z$ possesses KEKULÉ structures. This will occur if both the interior and the exterior parts of $B - Z$ possess KEKULÉ structures. If the size of Z is 4λ, then the interior part of $B - Z$ has an odd number of vertices and consequently cannot have KEKULÉ structures. This means that a cycle of size 4λ cannot be a conjugated circuit. Therefore, if Z is a conjugated circuit, its size must be $4\lambda + 2$.

Conjugated circuits in any benzenoid system are of the size 6, 10, 14, 18 etc. They cannot have size 12, 16, 20, 24 etc.

6.4 Connection with HERNDON's Resonance Theory

It has been pointed out many times that the conjugated circuit model to some extent is analogous with HERNDON's resonance theory. This is based on the fact that the coefficients ϱ_1 and ϱ_2 in (6.2) are exactly twice as large as the multipliers χ_1 and χ_2 in the HRT expression for the resonance energy, (5.14). This means that the choice $R_1 = 0.841$ eV, $R_2 = 0.336$ eV and $R_\lambda = 0$ for $\lambda > 2$ in formula (6.2) would yield resonance energies that are identical with the HRE values obtained from (5.14).

GOMES (1980, 1981) provided arguments in favour of the viewpoint that both the conjugated circuit model and HRT are oversimplified versions of a certain more realistic resonance theoretical approach.

It is easy to see why $2\varrho_1 = \chi_1$ and $2\varrho_2 = \chi_2$. As explained in Section 5.3, χ_λ counts the pairs of KEKULÉ structures which are transformed into each other by cyclically moving $2\lambda + 1$ double bonds. The possibility of cyclically moving $2\lambda + 1$ double bonds in a KEKULÉ structure means, however, that this KEKULÉ structure possesses a conjugated circuit of size $2 \times (2\lambda + 1) = 4\lambda + 2$. Thus χ_λ counts pairs of KEKULÉ structures possessing conjugated circuits of size $4\lambda + 2$. In other words, $2\chi_\lambda$ is equal to the number of $(4\lambda + 2)$-membered conjugated circuits in all KEKULÉ structures of the benzenoid molecule considered; both linearly independent and linearly dependent circuits are counted. Now, in RANDIĆ's theory all 6- and 10-membered conjugated circuits are linearly independent and therefore, for $\lambda = 1$ and $\lambda = 2$, ϱ_λ coincides with $2\chi_\lambda$. For larger values of λ this coincidence would be destroyed because in the HERNDON-type resonance theory all conjugated circuits are involved (GOMES 1980), whereas the RANDIĆ-type theories are restricted to linearly independent conjugated circuits.

SCHAAD and HESS (1982) have shown that if the linear independency restriction is abandoned (i.e. if all conjugated circuits are taken into consideration), then RANDIĆ's model becomes fully equivalent to HERNDON's resonance theory. Anyway, in recent applications of the conjugated circuit model (RANDIĆ, NIKOLIĆ, and TRINAJSTIĆ 1987, VOGLER and TRINAJSTIĆ 1988) linear independency of the conjugated circuits is no longer required.

From a practicioner's point of view it is sufficient to realize that both HERNDON's resonance theory and the conjugated circuit model give results (i.e. resonance energies) of similar accuracy and reliability. The conjugated circuit model is somewhat more flexible because it operates with four adjustable parameters, whereas HERNDON's theory requires only two such parameters. On the other hand, the application of the conjugated circuit model is usually quite difficult because of the necessity of recognizing linear dependency among conjugated circuits.

From a theoretician's point of view it must be embarrassing that more than ten years after the discovery of the conjugated circuit model no attempt has been made to give some plausible (preferably quantum-mechanical) explanation for its success. In particular, it is completely obscure what linear dependence between conjugated circuits might mean, and why (the greatest among) the linearly dependent circuits are to be disregarded.

6.5 Algorithms and Theorems

In order to apply formula (6.2) one has to find the numerical value of the four coefficients ϱ_1, ϱ_2, ϱ_3, ϱ_4. The first two of these coefficients can be calculated with the following simple formulas. As explained in the previous section, this is equivalent to finding the multipliers χ_1 and χ_2 in HERNDON's equation (5.14).

Consider a benzenoid system B. Denote by H a hexagon and by $B - H$ the system ($=$ graph) obtained by deleting from B the six vertices of H. Then

$$\varrho_1 = 2\chi_1 = 2 \sum_H K\{B - H\},$$

with the summation embracing all hexagons of B. A fully analogous identity holds for ϱ_2:

$$\varrho_2 = 2\chi_2 = 2 \sum_N K\{B - N\},$$

where N denotes a naphthalene subunit.

The above equations were proposed by HERNDON (1974b).

As an illustration, we compute ϱ_1 and ϱ_2 for benzo[a]anthracene. The necessary subgraphs are given as follows:

B − H$_1$;   K = 2

B − H$_2$;   K = 2

B − H$_3$;   K = 1

B − H$_4$;   K = 3

B − N$_1$;   K = 2

B − N$_2$;   K = 1

B − N$_3$;   K = 1

Consequently,

$$\varrho_1 = 2\chi_1 = 2 \times (2 + 2 + 1 + 3) = 16 \,,$$

$$\varrho_2 = 2\chi_2 = 2 \times (2 + 1 + 1) = 8 \,.$$

Formulas of this type for the third and fourth coefficient in Eq. (6.2) are significantly more complicated (GUTMAN 1990):

$$\varrho_3 = 2 \sum_{X\alpha} K\{B - X\alpha\} + \sum_{X\beta} K\{B - X\beta\} + 2 \sum_{X\gamma} K\{B - X\gamma\} \,,$$

$$\varrho_4 = 2 \sum_{Y\alpha} K\{B - Y\alpha\} + \sum_{Y\beta} K\{B - Y\beta\} + \sum_{Y\gamma} K\{B - Y\gamma\} + \sum_{Y\delta} K\{B - Y\delta\}$$

$$+ \sum_{Y\varepsilon} K\{B - Y\varepsilon\} + \sum_{Y\zeta} K\{B - Y\zeta\} + 2 \sum_{Y\eta} K\{B - Y\eta\} \,,$$

where $X\alpha$, $X\beta$, $X\gamma$ denote anthracene, phenanthrene and pyrene subunits, respectively, whereas $Y\alpha$, $Y\beta$, $Y\gamma$, $Y\delta$, $Y\varepsilon$, $Y\zeta$, $Y\eta$ denote naphthacene, benzo[a]anthracene, triphenylene, benzo[a]pyrene, benzo[e]pyrene, benzo[ghi]perylene and anthanthrene subunits, respectively.

When applying the conjugated circuit model it is very useful to know how many conjugated circuits are contained in the given KEKULÉ structure. Again, no simple rule exists for either the number of conjugated circuits or the number of independent conjugated circuits.

However, if also multiple conjugated circuits are included in the count, then this number is equal to the number of KEKULÉ structures minus one (GUTMAN and RANDIĆ 1979).

We have previously seen that the KEKULÉ structure 1 of benzo[a]pyrene contains 6 conjugated circuits. There are additional two doublets of nontouching conjugated circuits:

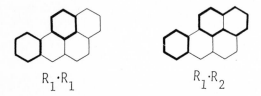

$$R_1 \cdot R_1 \qquad\qquad R_1 \cdot R_2$$

This gives a total of 8.

In the KEKULÉ structure 2, only five conjugated circuits were found. In this case there are three doublets:

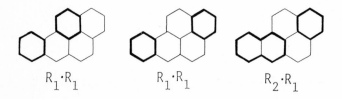

$$R_1 \cdot R_1 \qquad\qquad R_1 \cdot R_1 \qquad\qquad R_2 \cdot R_1$$

This again gives a total of 8. It is recalled that the KEKULÉ structure count of benzo-[*a*]pyrene is 9.

This interesting relation for the number of conjugated circuits is of very limited practical applicability, mainly because of the fact that the effect of multiple conjugated circuits (e.g. of $R_1 \cdot R_1$ or $R_1 \cdot R_2$) is not considered within the present version of the theory.

As has been pointed out several times, the application of the conjugated circuit model requires the inspection of each KEKULÉ structure and is therefore extremely difficult in the case of large benzenoid systems with hundreds and thousands of KEKULÉ structures.

Recently, an attractive solution of the problem was proposed (RANDIĆ, SOLOMON et al. 1987). Instead of them all, only a limited number of KEKULÉ structures are examined. These KEKULÉ structures are constructed at random and thus, according to laws of statistics, one can expect that they provide a representative sample. Then the circuit decomposition of these randomly selected KEKULÉ structures can be used to estimate the circuit decomposition of the benzenoid system considered, and thus its resonance energy.

Typical is the example of hexabenzo[*a, d, g, j, m, p*]coronene for which $K = 432$. RANDIĆ, SOLOMON, et al. (1987) constructed a random sample of 64 KEKULÉ structures. Their analysis gave $RE = 7.25$ eV, which was not far from the correct value $RE = 7.44$ eV.

Chapter 7

Aromatic Sextets

7.1 The History of the Aromatic Sextet

The exceptional stability of benzene has long puzzled the organic chemists, and several more or less plausible explanations of this fact have been offered. After the first rough ideas about the electronic structure of molecules had been achieved in the first decades of this century, it became clear that the stability of benzene can be associated with the occurrence of a stable electronic configuration which is a result of the cyclic arrangement of six electrons.

It seems that this point of view was first explicitly formulated by Sir ROBERT ROBINSON in an article which he published in 1925 together with J. W. ARMIT. The relevant quotation reads (ARMIT and ROBINSON 1925):

"The occurrence of many atoms in a molecule provides further opportunities for the emergence of electron groups of marked stability and, *ceteris paribus*, the possession of such groups confers chemical stability as shown, for example, by reduced unsaturation and a tendency to maintain the type. These are, of course, the chief characteristics of benzenoid systems, and here the explanation is obviously that six electrons are able to form a group which resists disruption, and may be called the *aromatic sextet*."

This aromatic sextet was then symbolically represented by inscribing a circle in the structural formula of benzene:

CLAR later qualified the use of the circle in the structural formulas of benzenoid hydrocarbons as "*one of the rare manifestations of chemical instinct of* KEKULÉ *style*" (CLAR 1972, p. 12).

In the words of ARMIT and ROBINSON (1925): "The circle in the ring symbolizes the view that six electrons in the benzene molecule produce a stable association which is responsible for the *aromatic* character of the substance."

It should be noted that in the same paper (ARMIT and ROBINSON 1925) the following structural formulas for naphthalene and anthracene were offered:

 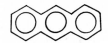

This can be viewed as an unsuccessful and exaggerated guess, which was not further pursued by organic chemists. Similarly, the frequently used formulas

are also unacceptable *if* one interprets the circle as a group of six electrons.

After a dormant period of more than 30 years, the idea of the aromatic sextet was revived by ERIC CLAR, who, in an article published in 1958 jointly with M. ZANDER, rationalized the differences in the behaviour of the two isomeric tribenzoperylenes $C_{30}H_{16}$:

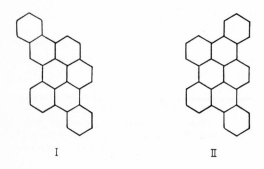

I   II

The authors observed that in I only four circles representing aromatic sextets can be drawn, whereas in II five such sextets can exist simultaneously:

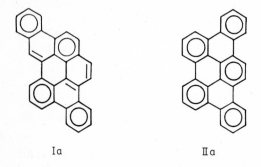

Ia   IIa

CLAR and ZANDER (1958) say:

"Unlike I which condensed quantitatively with maleic anhydride and chloranil, II did not react. The difference between the two hydrocarbons is that II can be considered to be a condensed quinquephenyl, as indicated in formula IIa, whilst no analogous arrangement in the isomer Ia is possible. In the latter at least three double bonds remain outside the benzenoid rings, thus making the molecule sufficiently reactive for the above condensation. According to formula II a, the molecule contains five benzenoid rings (each containing three double bonds) which are interlinked by quasi-single bonds, there being three quasi-empty rings. Such an arrangement accounts for the very low reactivity of the almost colourless hydrocarbon, which does not dissolve in cold concentrated sulphuric acid. Formula II a represents, not all possible arrangements of the double bonds, but the majority of all KEKULÉ structures, i.e. the rings indicated by circles are benzenoid."

These ideas led to the prediction that hexabenzocoronene III should be an unusually stable benzenoid hydrocarbon, because of its aromatic sextet formula III a. Eventually this prediction found a full experimental confirmation (CLAR et al. 1959).

III   III a

The concept of the aromatic sextet was later elaborated further by CLAR and outlined in full detail in his famous booklet (CLAR 1972). We present the basic features of CLAR's theory in the subsequent section.

A further major step forward was made in 1975 when HOSOYA and YAMAGUCHI introduced the concept of sextet polynomial. Their work revealed for the first time the algebraic and combinatorial contents of the aromatic sextet theory. In the later parts of the present chapter a detailed account of the research which followed along these lines is given.

7.2 CLAR's Aromatic Sextet Theory

7.2.1 CLAR Structural Formulas

The theory put forward by CLAR (1972) can be viewed as a consequent extension of the ideas of ARMIT and ROBINSON (1925) to the case of polycyclic systems. Whe-

reas the presentation of benzene in the form IV is more or less a matter of convention, in the case of

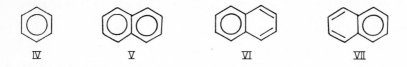

naphthalene we encounter some non-trivial problems. Formula V for naphthalene
(which can be found in many textbooks of organic chemistry) is in contradiction
with the original meaning of the circle, namely a group of six delocalized electrons.
Formula V would thus imply the existence of 12 delocalized electrons in naphthalene, whereas in reality the number of such electrons is only 10. In order to over-
come this difficulty CLAR associated with naphthalene the formula VI. Since there
is no particular reason to locate the circle in the left-hand side ring, another formula VII is equally plausible. CLAR proposed to describe naphthalene by the pair
of the formulas VI & VII and assumed some sort of migration of the sextet from
one ring into the other (CLAR 1972, p. 13). He expressed such a "sextet migration"
by formulas of the type VIII or IX:

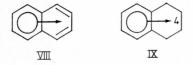

In the following, formulas of type VI–VIII will be called *Clar structural formulas*
or simply *Clar structures*.

The CLAR structures are constructed according to the following formal rules
(GUTMAN 1982a):
(a) Circles are never drawn in adjacent hexagons.
(b) The remainder of the benzenoid system, obtained by the deletion of the vertices
of the hexagons that possess circles, must have a KEKULÉ structure; this
remainder may be empty (in the case of all-benzenoid systems).
(c) As many circles as possible are drawn, subject to the constraints (a) and (b).

The three formulas of dibenzo[c, f]tetraphene below are not correct in the
sense of CLAR's theory because they violate conditions (a), (b), and (c), respectively.

Dibenzo[c, f]tetraphene has four CLAR structures, each containing three aromatic sextets and four double bonds:

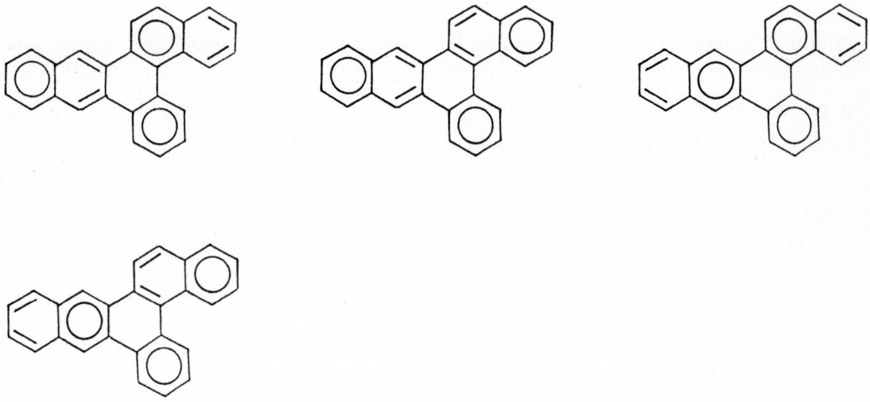

A shorthand notation for these four CLAR structures would be:

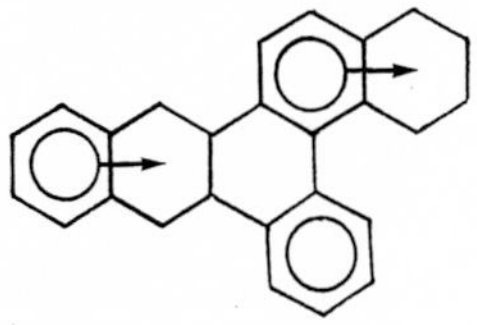

7.2.2 Some Chemical Implications of CLAR Structural Formulas

There are many benzenoid systems possessing a unique CLAR structural formula. In addition to II and III, phenanthrene may serve as an example of this kind:

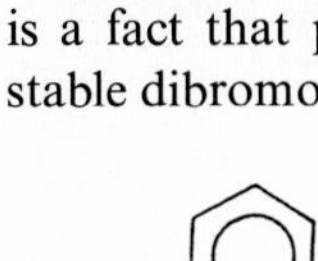

According to the theory of CLAR, phenanthrene is composed of two benzene-like regions and a double bond whose chemical character resembles that of olefins. It is a fact that phenanthrene easily adds bromine without a catalyst, forming a stable dibromo-derivative:

Within a series of isomeric benzenoid hydrocarbons the following general rules are observed. With increasing number of sextet

(a) the stability of the isomer increases, which means also lower chemical reactivity, and

(b) the absorption bands are shifted towards shorter wavelengths, which also means a change of color from dark blue-green to red, yellow or white.

A good example of this is provided by heptacene and its isomers; λ refers to the wavelength of the first absorption maximum in the p-band.

dark green, highly reactive, λ = ?

blue-green, λ = 651 nm

red, λ = 523 nm

violet-red, λ = 538 nm

yellow, λ = 423 nm

white, very stable, λ = 334 nm

Observe that dibenzo[a, l]pentacene and dibenzo[a, c]pentacene from the above examples, which both have three aromatic sextets, also have quite similar color, since their absorption spectra almost coincide. Their chemical properties are also similar.

CLAR offered a vast amount of experimental data supporting his theory (CLAR 1972). These include absorption spectra, NMR coupling constants and, principally, numerous characteristics of the chemical reactivity of benzenoid hydrocarbons.

Here we mention as a typical example the CH_3 doublet separation in the NMR spectrum of the two isomeric dimethyl-benzoperylenes X and XI.

In X the CH_3 signals are separated by 0.6 Hz, which is the same value as found in dimethylperylene (XII). In compound XI the CH_3 doublet separation is 1.1 Hz. This shows that the methyl groups in XI are attached to a localized double bond, contrary to X and XII.

7.2.3 All-Benzenoid Systems

A distinguished role in CLAR's theory is played by those benzenoid hydrocarbons whose (unique) CLAR structure does not possess double bonds. They are called all-benzenoid hydrocarbons. (CLAR used the name fully benzenoid hydrocarbons.)

We have already met a few such benzenoid systems. Here are some further examples:

The hexagons of an all-benzenoid system are in a natural way divided into full (which possess a circle in the CLAR structure) and empty (in which no circles are drawn). It is not difficult to see that each empty hexagon must have exactly three full neighbors. Furthermore, the number of vertices of an all-benzenoid system is necessarily divisible by 6. More details about the structure of all-benzenoids can be found elsewhere (POLANSKY and ROUVRAY 1976b, GUTMAN and CYVIN 1988b). A number of results concerning the enumeration of all-benzenoid systems is presented in Chapter 4.

All-benzenoid hydrocarbons exhibit a pronounced chemical inertness. They are usually pale yellow or colorless solids. Contrary to other benzenoid hydrocarbons they do not dissolve in cold concentrated sulphuric acid. They do not react with maleic anhydride. Samples of some all-benzenoid hydrocarbons, prepared by CLAR about fifty years ago, show no sign of decomposition, although they are kept at room temperature, exposed to air and light.

The stability of all-benzenoid hydrocarbons is well illustrated by the fact that the melting point of hexabenzocoronene (III) could not be determined because the glass tube used in the experiment melted long before the hydrocarbon (CLAR 1972, p. 28).

7.2.4 CLAR Structures and KEKULÉ structures

CLAR structural formulas are traditionally viewed as a shorthand representation of some of the KEKULÉ structures of the respective benzenoid system (e.g. see CLAR 1972, p. 29). Thus the CLAR structure of pyrene is assumed to represent 4 KEKULÉ structures. Two more KEKULÉ structures are not accounted for. In CLAR's theory they are usually considered to be less important.

The situation may also be the reverse. It may happen that one and the same KEKULÉ structure is represented by several CLAR structures. This is illustrated by a

The Clar structure of pyrene and the four Kekulé structures
compatible with it

Kekulé structures of pyrene not "contained"
in the Clar structure

KEKULÉ structure of bisanthene. Recall that bisanthene has 16 KEKULÉ and 9 CLAR structures.

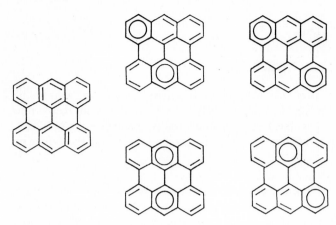

A Kekulé structure of bisanthene compatible with four
Clar structures

The fractions of those KEKULÉ structures which are represented by the CLAR structures of a benzenoid system can differ considerably. There are benzenoids in which all KEKULÉ structures are "contained" in CLAR structures. In other cases, however, the number of such KEKULÉ structures may be very small compared to the total KEKULÉ structure count, and can be made negligibly small with the increasing size of the molecule (GUTMAN, OBENLAND, and SCHMIDT 1985).

As a conclusion we may say that it is not always true that CLAR structures account for the majority of the KEKULÉ structures. It would be quite risky to disregard those, sometimes quite numerous, KEKULÉ structures which are not compatible with a CLAR structural formula.

*

In this section we gave only a brief outline of the basic features of CLAR's theory. Those interested in its details should obligatorily consult CLAR's original work (CLAR 1972).

*

After the concept of the CLAR structure was developed, several attempts were made to justify and rationalize it quantum mechanically. Various approaches have been employed (PAUNCZ and COHEN 1960, POLANSKY and DERFLINGER 1967, RANDIĆ 1974, GUTMAN, and BOSANAC 1977, AIDA and HOSOYA 1981). The general conclusion of these researches is that in the majority of cases the CLAR formulas are in harmony with the electronic structure of benzenoid molecules as determined by quantum-chemical methods. Hexagons in which circles are drawn correspond to regions in the molecule in which a high π-electron density is found. Empty hexagons correspond to regions with low π-electron density. Such regularities are particularly pronounced in benzenoids for which the CLAR structure is unique (all-benzenoid systems).

A detailed discussion of this matter goes beyond the scope of the present book.

7.3 Generalized CLAR Structures

In 1975 HOSOYA and YAMAGUCHI made a significant step forward in the aromatic sextet theory by considering CLAR-type structures in which it was not required to have as many sextets as possible. Bearing in mind the discussion in Paragraph 7.2.1, the generalized CLAR structures can be constructed according to the following formal rules:

(a) Circles are never drawn in adjacent hexagons.
(b) The remainder of the benzenoid system, obtained by deletion of the vertices of the hexagons which possess circles, must have a KEKULÉ structure; this remainder may be empty (in the case of the CLAR structure of an all-benzenoid system).

Each CLAR structure is also a generalized CLAR structure.

In the case of generalized CLAR structures the double bonds need not have a unique position and are therefore not indicated. By definition, there exists a unique generalized CLAR structure with no aromatic sextets. We shall call this formal extension of the CLAR-structure concept the *trivial generalized* CLAR *structure*.

The two formulas below of dibenzo[c, f]tetraphene are not correct in the sense of the above definition because they violate conditions (a) and (b), respectively.

Dibenzo[c, f]tetraphene has a total of 20 generalized CLAR structures of which we present four. They contain 1, 2, 2, and 3 sextets, respectively.

The trivial generalized CLAR structure is simply drawn with all hexagons empty:

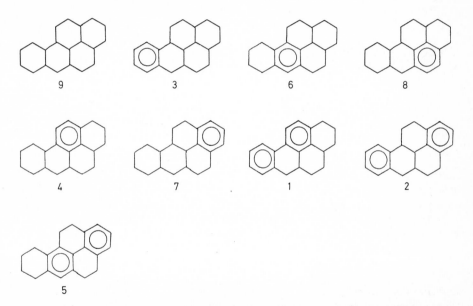

the trivial generalized Clar structure
of dibenzo[c,f]tetraphene

In Fig. 7.1 is depicted the complete set of generalized CLAR formulas of benzo[a]-pyrene.

Fig. 7.1. The generalized CLAR structures of benzo[a]pyrene. Their number coincides with the number of KEKULÉ structures. The numbering of the generalized CLAR structures is made so as to match with the KEKULÉ structures depicted in Fig. 5.1; for explanation see text. No. 9 is the trivial generalized CLAR structure

The generalization of the CLAR-structure concept described above is in drastic dissonance with the original ideas of CLAR's theory. Furthermore, the chemical meaning of the generalized CLAR structures (and in particular of the trivial structure) is to a great extent obscure. As a matter of fact, no attempt has ever been made to find a plausible interpretation.

Nevertheless, the new concept is not completely meaningless, as can be seen from the fact that (in many cases) the number of generalized CLAR structures is equal to the number of KEKULÉ structures.

This remarkable result was observed first by HOSOYA and YAMAGUCHI (1975) and was eventually rationalized by the discovery of a one-to-one mapping between generalized CLAR and KEKULÉ structures (OHKAMI and HOSOYA 1983). The mapping assumes the replacement of three double bonds in the arrangement A by a circle. In all other cases, including also the arrangement B, the hexagon is left empty.

A   B

Thus, for instance, the KEKULÉ and generalized CLAR structures below of dibenzo[c, f]tetraphene are mapped onto each other.

There is a KEKULÉ structure of dibenzo[c, f]tetraphene in which no hexagon has three double bonds arranged as A; it is mapped onto the trivial generalized CLAR structure.

The one-to-one correspondence between the nine KEKULÉ and the nine generalized CLAR structures of benzo[a]pyrene can be seen by comparing Figs. 5.1 and 7.1.

The regularity described above does not hold, unfortunately, for all benzenoid systems. HOSOYA and YAMAGUCHI (1975) observed that coronene has 20 KEKULÉ structures, but only 19 generalized CLAR structures. ZHANG and CHEN (1986) clarified this point by proving a result to be stated shortly. To formulate it we need some preparation.

Denote coronene by C. If B is a benzenoid system possessing a coronene sub-unit, then $B - C$ denotes the remainder obtained by deleting the vertices of C from B. An example is given below.

C   B   B - C

Let $K\{B\}$, as before, denote the number of KEKULÉ structures of B. Then the theorem of ZHANG and CHEN is as follows:

A benzenoid system B has an equal number of generalized CLAR and KEKULÉ structures if and only if

(a) *B possesses no coronene subunit, or*
(b) *B is not coronene, or*
(c) *for each coronene subunit C contained in B, $K\{B - C\} = 0$.*

In the above example $K\{B - C\} = 6$. Therefore in this case there is no one-to-one mapping between the generalized CLAR and the KEKULÉ structures.

Note that the ZHANG-CHEN theorem guarantees that the one-to-one mapping between generalized CLAR and KEKULÉ structures exists for all catacondensed systems; this special case was proved somewhat earlier by GUTMAN, HOSOYA et al. (1977).

7.3.1 The Super-Ring

From the fact that coronene possesses 20 KEKULÉ structures, but only 19 generalized CLAR structures, one immediately arrives at the idea that the CLAR structure concept should be further extended. HOSOYA and YAMAGUCHI (1975) proposed the super-ring, that is, an additional generalized CLAR structure, which, instead of a circle (representing a sextet of electrons), contains an 18-membered cycle – the perimeter of coronene (representing thus a group of 18 electrons). The respective additional CLAR-type formula for coronene would be of the following form:

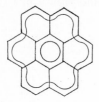

With this extension the one-to-one correspondence between CLAR and KEKULÉ structures was expected to be maintained in the general case (HOSOYA and YAMA-GUCHI 1975, OHKAMI and HOSOYA 1983). This is indeed the case as was recently demonstrated by HE and HE (1987).

The details of the super-ring theory will not be outlined in the present book; they can be found in the works of OHKAMI and HOSOYA (1983) and HE and HE (1986, 1987).

7.4 The Sextet Polynomial

The sextet polynomial is the first genuine mathematical object introduced within the aromatic sextet theory (HOSOYA and YAMAGUCHI 1975). It is defined in the following manner.

Let B be a benzenoid system. Denote by $s(B, k)$ the number of generalized CLAR structures of B having k aromatic sextets. Denote by S the largest number of sextets which can be drawn in B, this means that $s(B, S)$ is the number of (nongeneralized) CLAR structures.

Then the polynomial in the variable x

$$\sigma(B, x) = s(B, 0) + s(B, 1)\, x + s(B, 2)\, x^2 + \ldots + s(B, S)\, x^S$$

or, what is the same,

$$\sigma(B, x) = \sum_{k=0}^{S} s(B, k)\, x^k$$

is called the *sextet polynomial* of the benzenoid system B.

As an illustration we calculate the sextet polynomial of benzo[a]pyrene. The respective generalized CLAR structures are given in Fig. 7.1. From Fig. 7.1 we easily see that for $B = $ benzo[a]pyrene

$s(B, 0) = 1$ (structure 9),
$s(B, 1) = 5$ (structures 3, 4, 6, 7, 8),
$s(B, 2) = 3$ (structures 1, 2, 5),

and $S = 2$. This means that for $k > 2$, $s(B, k) = 0$. The sextet polynomial of benzo[a]pyrene reads:

$$\sigma(B, x) = 1 + 5x + 3x^2 .$$

At this point it would be instructive for the reader to verify that the sextet polynomial of coronene C is

$$\sigma(C, x) = 1 + 7x + 9x^2 + 2x^3 .$$

Note that, by definition, there is just one trivial generalized CLAR structure; this means that for all B, $s(B, 0) = 1$, and all sextet polynomials have the form $\sigma(B, x) = 1 +$ terms depending on x.

The following property of the sextet polynomial

$$\sigma(B, 1) = K\{B\} \tag{7.1}$$

is just another way of expressing the one-to-one correspondence between KEKULÉ and generalized CLAR structures. It is not generally valid, as discussed in detail in the preceding section. Coronene is an example where (7.1) is not satisfied.

Another, less obvious, relation for the sextet polynomial reads:

$$\sigma'(B, 1) = \chi_1 = \frac{1}{2} \varrho_1 , \tag{7.2}$$

where χ_1 is the resonance-theoretical parameter from (5.14), whereas ϱ_1 is used in the conjugated circuit model, (6.2); σ' symbolizes the first derivative of σ.

Formula (7.2) has the same range of validity as (7.1). It was first observed by HOSOYA and YAMAGUCHI (1975). Its applicability to all catacondensed benzenoids was proved by GUTMAN (1981).

A third, nontrivial algebraic property of $\sigma(B, x)$ is the following result, whose origin will become obvious in the subsequent section.

If B is an unbranched catacondensed benzenoid system, then all roots of the equation $\sigma(B, x) = 0$ are real negative numbers.

7.5 Concepts Related to the Sextet Polynomial

The pioneering paper of HOSOYA and YAMAGUCHI (1975) stimulated a great number of algebraic and combinatorial approaches to CLAR's theory or its appropriate generalizations. A variety of new concepts has been introduced. Some of them are briefly outlined in this section.

7.5.1 AIHARA's Resonance Energy

AIHARA (1977) examined the following modified form of the sextet polynomial:

$$\sigma^*(B, x) = \sum_{k=0}^{S} (-1)^k s(B, k) x^{2(S-k)},$$

and denoted by $x_1, x_2, \ldots, x_{2S}$ the roots of the equation $\sigma^*(B, x) = 0$. If these roots are real numbers (i.e. not complex numbers), then half of them (say, x_1, $x_2, \ldots, x_S$) are positive and half of them negative (say, $x_{S+1}, \ldots, x_{2S}$). AIHARA (1977) proposed the quantity

$$RE^*(B) = \sum_{i=1}^{S} x_i \qquad\qquad (7.3)$$

as a measure of the degree of aromaticity of the corresponding benzenoid molecule B.

For example, if $B = $ benzo[a]pyrene, then (using the results obtained in the previous section)

$$\sigma^*(B, x) = x^4 - 5x^2 + 3,$$

and it is easily computed that

$$x_1 = \sqrt{(5 + \sqrt{13})/2} = 2.074,$$
$$x_2 = \sqrt{(5 - \sqrt{13})/2} = 0.835,$$
$$x_3 = -x_2; \qquad x_4 = -x_1,$$

resulting in

$$RE^* \text{ (benzo[a]pyrene)} = 2.909.$$

It has been pointed out that $RE^*(B)$ is roughly proportional to $\ln K\{B\}$ (AIHARA 1977, GUTMAN 1978), a result which should be compared with (5.15). Extensive calculations (GUTMAN and EL-BASIL 1985) showed that the stability of benzenoid hydrocarbons predicted on the basis of their RE^* values often disagrees with the conclusions drawn from other theoretical approaches.

Another difficulty in the RE^* method is that the roots of $\sigma^*(B, x) = 0$ may in some cases be complex numbers. Then RE^*, as defined by means of (7.3), would be complex-valued too, and thus without physical meaning.

7.5.2 The CLAR Graph

A hexagon χ of a benzenoid system is said to be *resonant* if there is a generalized CLAR structure having a circle in χ. Otherwise, χ is *nonresonant* (HOSOYA and YAMAGUCHI 1975).

For instance, the hexagons 1, 2, 3, 6, 7 and 8 of bisanthene are resonant whereas the hexagons 4 and 5 are nonresonant.

There are many benzenoids all of whose hexagons are resonant (e.g. all cata-condensed systems).

Two hexagons χ_i and χ_j are said to be *mutually resonant* if there is a generalized CLAR structure having circles in both χ_i and χ_j. In the above example (1,6), (1,7), (2,7), etc. are pairs of mutually resonant hexagons. On the other hand (1,2), (1,3), (1,5), etc. are pairs of hexagons which are not mutually resonant.

Note that adjacent hexagons cannot be mutually resonant.

Let B be a benzenoid system with hexagons $\chi_1, \chi_2, \dots, \chi_h$. Then the CLAR *graph* $C(B)$ of B is defined as follows (GUTMAN 1982b): $C(B)$ has h vertices χ_1, $\chi_2, \dots, \chi_h$. The vertices χ_i and χ_j of $C(B)$ are adjacent if the hexagons χ_i and χ_j in B are not mutually resonant.

Below are presented the CLAR graphs of benzo[a]pyrene and dibenzo[a, n]pery-lene. From these examples the construction of the CLAR graph should become obvious.

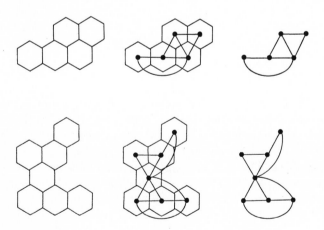

Two benzenoid systems and their Clar graphs

One should observe that if χ is a non-resonant hexagon of a benzenoid system B then the vertex corresponding to χ in $C(B)$ is adjacent to all other vertices of $C(B)$.

The idea behind the definition of the CLAR graph lies in the following. Let G be a graph and denote by $n(G, k)$ the number of ways in which k mutually non-adjacent vertices can be selected in G. By definition, $n(G, 0) = 1$ and $n(G, 1) = \#$ vertices of G. The polynomial

$$\omega(G, x) = \sum_k n(G, k)\, x^k$$

is called the *independence polynomial* of the graph G (for details, see: CVETKOVIĆ et al. 1988).

The following identities are obeyed (GUTMAN 1982b):

$$s(B, k) = n(C(B), k)\,, \qquad \text{for} \qquad k = 0 \qquad \text{and} \qquad k > 1\,,$$

$$s(B, 1) = n(C(B), 1) - \#\text{ nonresonant hexagons}\,,$$

resulting in

$$\sigma(B, x) = \omega(C(B), x) - x \times \#\text{ nonresonant hexagons}\,. \tag{7.4}$$

In the often occuring case when all hexagons of B are resonant, (7.4) reduces to

$$\sigma(B, x) = \omega(C(B), x)\,.$$

This latter equality holds, for example, for all catacondensed systems.

The usefulness of relation (7.4) lies in the fact that there exist simple and efficient procedures for the calculation of the independence polynomial; see the subsequent paragraph. Since the number of nonresonant hexagons is easily established by inspecting the benzenoid system, (7.4) provides an economic route for the calculation of the sextet polynomial (GUTMAN 1982b, GUTMAN and EL-BASIL 1984).

Equation (7.4) holds for all benzenoid systems.

The CLAR-graph concept eventually found quite a few additional applications (HERNDON and HOSOYA 1984, EL-BASIL 1986a, EL-BASIL and RANDIĆ 1987). In connection with it, the CLAR *matrix* has been defined as a square matrix of order h whose (ij)-entry is equal to 1 if the hexagons χ_i and χ_j are not mutually resonant and is equal to 0 otherwise (EL-BASIL 1986a). Evidently, the CLAR matrix is just the adjacency matrix of the CLAR graph.

The CLAR *coloring* (EL-BASIL 1986a) is coloring of the vertices of the CLAR graph with two colors (black and white) so that no two black vertices are adjacent and that every white vertex has a black neighbor. Let the number of different CLAR colorings of $C(B)$ such that k vertices are black be denoted by $c(B, k)$. Then the polynomial

$$\gamma(B, x) = \sum_k c(B, k)\, x^k$$

has been named the CLAR *polynomial* of the benzenoid system B (EL-BASIL 1986a). CLAR colorings and CLAR polynomials play some role in the valence-bond method of HERNDON and HOSOYA (see Section 7.6), but, until now, only a few of their properties have been established (EL-BASIL and RANDIĆ 1987).

7.5.3 Calculation of the Independence Polynomial

Let G be a graph, v its vertex and A_v the collection of vertices containing v and its first neighbors. Then the independence polynomial of G satisfies the recurrence relation

$$\omega(G) = \omega(G - v) + x\omega(G - A_v) .$$

Example:

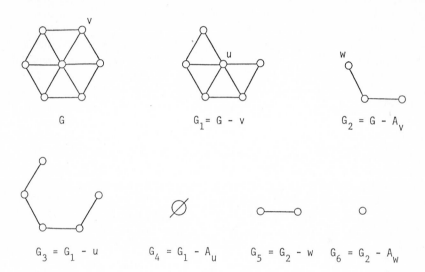

From the definition of the independence polynomial we immediately have

$$\omega(G_4) = 1, \qquad \omega(G_5) = 1 + x, \qquad \omega(G_6) = 1 + 2x ,$$

whereas the derivation of

$$\omega(G_3) = 1 + 5x + 6x^2 + x^3$$

is left to the reader as an easy exercise. Therefore

$$\omega(G_1) = (1 + 5x + 6x^2 + x^3) + x(1) = 1 + 6x + 6x^2 + x^3 ,$$
$$\omega(G_2) = (1 + 2x) + x(1 + x) = 1 + 3x + x^2 ,$$
$$\omega(G) = (1 + 6x + 6x^2 + x^3) + x(1 + 3x + x^2)$$
$$= 1 + 7x + 9x^2 + 2x^3 .$$

Since G is the CLAR graph of coronene, we have just computed the sextet polynomial of coronene. Note also that G_3 is the CLAR graph of picene.

Another relation useful for the calculation of the independence polynomial is
the following. If the graph G is composed of two disconnected parts G_a and G_b,
then

$$\omega(G) = \omega(G_a) \, \omega(G_b) \,.$$

7.5.4 The GUTMAN Tree

The CLAR graph is constructed in such a way that every *selection of mutually non-
adjacent vertices* in it represents a generalized CLAR structure of the corresponding
benzenoid system. By means of the CLAR graph the computation of the sextet
polynomial is reduced to the finding of an independence polynomial. This, as was
shown in the preceding section, is quite easy.

In the case of unbranched catacondensed systems it is possible to construct
another graph, such that every *selection of mutually nonincident edges* in it represents a generalized CLAR structure (GUTMAN 1977b). This graph is connected and
acyclic, and is therefore in the standard terminology of graph theory called a tree.
This graph has been namend the GUTMAN *tree* (EL-BASIL 1984a, 1984b, 1986b,
1987); another name for it is *caterpillar* or *caterpillar tree* (HARARY and SCHWENK
1973).

A hexagon in an unbranched catacondensed benzenoid system is of the mode
L_1, L_2, or A_2 (see Section 3.2 and Fig. 3.2). Exactly two hexagons have the mode L_1.

Let B be an unbranched catacondensed benzenoid system whose hexagons
$\chi_1, \chi_2, \dots, \chi_h$ are labeled so that χ_i is adjacent to $\chi_{i+1}, i = 1, 2, \dots, h-1$. The
LA-sequence of B is a sequence of symbols L and A such that the i-th member in
the sequence is L if χ_i is of the mode L_1 or L_2, and is A if χ_i is of the mode A_2.

An example is given below.

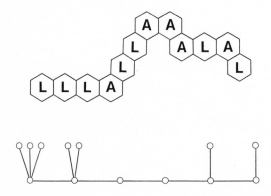

An unbranched catacondensed benzenoid system whose LA-sequence

is $LLLALLAAALAL$, and the associated Gutman tree

Using the conventions $LL = L^2$, $LLL = L^3$, etc., and $L^0 =$ no symbol L, the LA-sequence of B can be presented in the form

$$L^{t_0} AL^{t_1} AL^{t_2} \dots AL^{t_p} , \qquad (7.5)$$

where $t_i \geq 0$ for $i = 1, 2, \dots, p-1$, $t_0 > 0$, $t_p > 0$. The caterpillar tree corresponding to the unbranched catacondensed benzenoid system B whose LA-sequence is given by (7.5) has the following structure (GUTMAN 1977b).

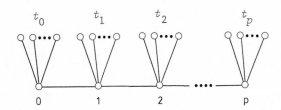

Gutman tree associated with

the LA-sequence (7.5)

If $T = T(B)$ is the caterpillar tree associated with B, then the identities

$$s(B, k) = m(T, k)$$

are obeyed for all $k = 0, 1, \dots, S$, where $m(T, k)$ is the number of ways in which one can select k mutually nonincident edges in T. The polynomial $\alpha(T, x)$, viz.

$$\alpha(T, x) = \sum_k m(T, k) \, x^k ,$$

is the *matching polynomial* of the graph T (LOVÁSZ and PLUMMER 1986, CVETKOVIĆ et al. 1988)*. Consequently,

$$\sigma(B, x) = \alpha(T, x) . \qquad (7.6)$$

The most obvious application of the identity (7.6) is for the calculation of the sextet polynomial. The matching polynomial of a tree can be easily calculated, especially

---

* For the matching polynomial it is essential that its coefficients are the numbers $m(G, k)$. The most often employed form of the matching polynomial is

$$\sum_k (-1)^k \, m(G, k) \, x^{n-2k} ,$$

where n is the number of vertices of the graph G. We use here the form which is more convenient for the aromatic sextet theory. Note that the matching polynomial has numerous applications and that its mathematical theory is well elaborated (LOVÁSZ and PLUMMER 1986, CVETKOVIĆ et al. 1988).

if one has in mind its relation to the characteristic polynomial. Since we have already presented an easy method for the computation of $\sigma(B, x)$, which is not limited to unbranched catacondensed systems, we will mention only some other implications of (7.6).

First of all, (7.6) guarantees that the roots of the equation $\sigma(B, x) = 0$ are real numbers. That is, the zeros of matching polynomials are known always to be real. Consequently, the AIHARA resonance energy, (7.3), is real-valued, at least in the case of unbranched catacondensed benzenoids.

Further, by means of (7.1), which certainly holds for unbranched catacondensed systems, the unexpected connection

$$K\{B\} = \sum_k m(T, k) = Z(T)$$

is obtained, where $Z(T)$ is HOSOYA's topological index, originally designed for the study of the thermodynamic properties of alkanes (HOSOYA 1971).

A large number of additional relations which exist between the caterpillar trees and other combinatorial objects (rook boards, king boards, YOUNG diagrams, etc.) have been noticed recently (EL-BASIL 1987) and many novel combinatorial features of CLAR theory pointed out (EL-BASIL 1988).

7.6 The Valence-Bond Method of HERNDON and HOSOYA

HERNDON and HOSOYA (1984) proposed a method for the calculation of the resonance energy of benzenoid hydrocarbons based on CLAR structures. This is one of the few *quantitative* applications of the CLAR-type concepts and the only attempt to use CLAR structures as a basis set for *quantum chemical* calculations.

$H_{11} = 3A + 3B$

$H_{22} = 3A + 3B$

$H_{33} = 2A + 6B$

$H_{44} = 2A + 6B$

$H_{55} = 2A + 6B$

$H_{66} = A + 9B$

The basis set in the HERNDON-HOSOYA theory is formed by those generalized CLAR formulas to which no more circles can be added without violating the conditions (a) and (b) stated in Section 7.3.

In the case of benzo[a]pyrene, Fig. 7.1, the basis set is formed by the generalized CLAR structures 1, 2, 5, and 8. The basis set of coronene consists of six structures depicted above. For reasons which will become clear in a while, the double bonds are also indicated (although their position is not always unique).

Using such an N-dimensional basis set the $N \times N$ Hamiltonian matrix H is constructed according to the following rules.

The diagonal elements H_{ii} are equal to $a_iA + b_iB$ where a_i is the number of resonant sextets and b_i the number of double bonds in the i-th CLAR structure. A and B are considered the COULOMB integrals of an aromatic sextet and a double bond, respectively, but in reality they are semiempirical parameters. An illustration was given above.

Fixed double bonds are not counted in b_i.

The off-diagonal elements of H are equal to C, $C/2$, $C/4$ or 0, depending on the number of aromatic sextets which have to migrate to transform one CLAR structure into the other. C is a semiempirical parameter.

If the i-th CLAR structure is transformed into the j-th by the migration of 1, 2 and 3 aromatic sextets, then H_{ij} is equal to C, $C/2$ and $C/4$, respectively. Otherwise, H_{ij} is equal to zero. If the migrating sextets are separated by fixed single bonds, then $H_{ij} = 0$.

Detailed rules for counting the migrating sextets have been elaborated (HERNDON and HOSOYA 1984).

The Hamiltonian matrix of coronene has finally the following form:

$$H(\text{coronene}) = \begin{bmatrix} 3A+3B & C/4 & C & C & C & C \\ C/4 & 3A+3B & C & C & C & C \\ C & C & 2A+6B & C/2 & C/2 & 0 \\ C & C & C/2 & 2A+6B & C/2 & 0 \\ C & C & C/2 & C/2 & 2A+6B & 0 \\ C & C & 0 & 0 & 0 & A+9B \end{bmatrix}$$

Using the usual procedure of valence-bond theory, the matrix H is diagonalized. The eigenvalue associated with the totally symmetric eigenvector is the ground state energy, which, within the HERNDON–HOSOYA method, is interpreted as the resonance energy. Another assumption is that the CLAR structures provide an orthogonal basis set.

The parameters A, B and C were estimated by least-squares fitting:

$$\left. \begin{array}{l} A = 0.8590 \\ B = 0.0744 \\ C = 0.3176 \end{array} \right\} \begin{array}{l} \text{adjusted for resonance} \\ \text{energies in eV} \end{array}$$

With these values of A, B and C, the resonance energies of DEWAR and DE LLANO (1969) could be reproduced with an average deviation of only 0.013 eV and a correlation coefficient of 0.9998.

*

The CLAR coloring defined in Paragraph 7.5.2 is a graph-theoretical equivalent of the selection of the basis set in the HERNDON-HOSOYA theory. Every black vertex corresponds to a hexagon in which a circle has been drawn. The requirement that every white vertex has a black neighbor implies that no further circles can be added.

The k-th coefficient $c(B, k)$ of the CLAR polynomial is just the number of HERNDON-HOSOYA basis functions with k aromatic sextets. Consequently, the value of the CLAR polynomial $\gamma(B, x)$ for $x = 1$ is equal to the size of the basis set.

Chapter 8

Coronoids

In this chapter we outline the theory of coronoid hydrocarbons (chemical objects) and the related coronoid systems (mathematical objects). As it will be explained in more detail in the following, the coronoids can be viewed as a sort of benzenoids with holes. Thus the class of coronoids represents a proper extension of the class of benzenoids.

However, some significant distinctions between these two classes as well as between the corresponding chemical theories should be pointed out. Whereas the benzenoid molecules attracted the attention of theoretical chemists over a very long period of time, the study of coronoid molecules has taken place during the few last decades. Whereas several hundreds of benzenoid hydrocarbons are known, only two coronoid hydrocarbons have been synthesized so far. And whereas some benzenoid hydrocarbons (and their derivatives) have a great importance for the chemical and pharmaceutical industry, it is too early to judge about possible practical applications of the coronoid compounds.

Most of the present theoretical investigations of coronoid systems can be grouped into

(a) enumeration of coronoid systems, and
(b) enumeration of KEKULÉ structures in coronoids.

Both fields of research are covered in sufficient detail in the later parts of this chapter.

Finally, we wish to call the reader's attention to the fact that a number of elegant and powerful statements valid for benzenoid systems do not hold for coronoids. As examples we mention that both the DEWAR-LONGUET-HIGGINS formula (Paragraph 5.1.5) and the JOHN-SACHS formula (Paragraph 5.1.6) fail for certain coronoid systems. These features may be viewed as a challenge for mathematical chemists.

8.1 Coronoid Hydrocarbons

Like the benzenoid hydrocarbons (Chapter 1) coronoid hydrocarbons are condensed polycyclic molecules composed exclusively of six-membered rings. Also, coronoid hydrocarbons are unsaturated fully conjugated molecules.

The only few coronoid hydrocarbons which have been synthesized belong to a class called *cycloarenes* (STAAB and DIEDERICH 1983). These compounds are catacondensed rings forming a macrocyclic system so that a cavity is present, into which carbon-hydrogen bonds point. Below we show some clarifying examples.

C_1 C_1

C_2 C_3

The depicted molecules are $C_{48}H_{24}$, cyclo[*d.e.d.e.d.e.d.e.d.e.d.e*]dodecakisbenzene (C_1), $C_{40}H_{20}$, cyclo[*d.e.d.e.e.d.e.d.e.e*]decakisbenzene (C_2) and .$C_{36}H_{18}$, cyclo-[*d.e.e.d.e.e.d.e.e*]nonakisbenzene (C_3). In this nomenclature the symbols *d* and *e* [in brackets] reflect the linear and angular annelation of a benzenoid ring, respectively.

The first report on the attempts to synthesize C_1 was given in 1965 by H. A. STAAB at the Annual Meeting of ,,Gesellschaft Deutscher Chemiker" in Bonn, the KEKULÉ Centennial. On this occasion the molecule C_1 was given the name kekulene. Not until thirteen years later (DIEDERICH and STAAB 1978) was the first successful synthesis of kekulene reported. It turned out to be a greenish-yellow microcrystalline substance with an extreme insolubility in solvents of all kinds. Eight years after kekulene the second cycloarene C_2, was prepared (FUNHOFF and STAAB 1986). A synthesis of C_3 is also under way (STAAB and SAUER 1984).

8.2 Coronoid Systems

A coronoid system is a planar geometric figure that, like a benzenoid system, consists of congruent regular hexagons. Loosely speaking, a coronoid system is a benzenoid-like system with a hole; cf. the end of Chapter 2.

The definition of a benzenoid system in terms of a cycle on the hexagonal lattice (Definition C of Chapter 2) is easily adapted to a coronoid system. Assume two cycles, C' and C'', where C'' is completely embraced by C'. The vertices and edges on C'' and its interior are referred to as the *corona hole*. The size of C'' must be greater than 6. A coronoid system consists of the vertices and edges on C' and C'' as well as in the interior of C', but outside C''.

The cycles C' and C'' are the *outer perimeter* and the *inner perimeter* of the coronoid system, respectively. An illustration of this definition of coronoid systems is given below.

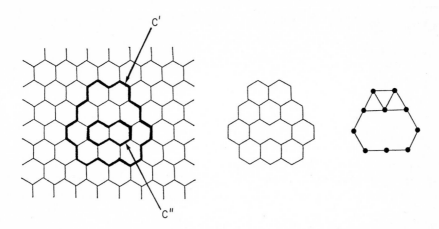

The above diagrams exemplify also the representation of a coronoid system by means of the dualist graph.

The present definition of coronoid systems applies, strictly speaking, only to *single coronoid systems*, i.e. those with exactly one hole. Throughout this chapter, only single coronoid systems are considered, and it is tacitly assumed that the term "coronoid" actually refers to "single coronoid".

In order to prepare for an alternative definition of coronoid systems we introduce the class of primitive coronoid systems.

A *primitive coronoid system* consists of a single chain of hexagons in a macrocyclic arrangement. The coronoid systems corresponding to C_1, C_2, and C_3 (Section 8.1) are all examples of primitive coronoid systems.

Now, it is inferred that all coronoid systems with $h + 1$ hexagons are constructed by

(a) adding one hexagon to a coronoid system with h hexagons, and
(b) including the primitive coronoids with $h + 1$ hexagons.

The addition of hexagons is taken in the same sense as in the corresponding construction of benzenoids (see Definition B of Chapter 2). It is emphasized that allowance is made for an addition to the inner perimeter (as well as to the outer), if such an addition is possible.

The above characterization is not an effective definition unless we have a way to construct all primitive coronoid systems for a given number of hexagons. This is achieved by allowing for an addition of a hexagon to a single (catacondensed) unbranched chain so that the chain becomes closed. This type of addition is re-referred to as *corona-condensation* (POLANSKY and ROUVRAY 1977).

The term "coronoid" was introduced by BRUNVOLL, CYVIN and CYVIN (1987a). More or less synonymous terms which have been used are:

coronaphene	(JENNY and PETER 1965, BALABAN and HARARY 1968)
corona-condensed system	(BALABAN and HARARY 1968, POLANSKY and ROUVRAY 1977)
(true) circulene	(DOPPER and WYNBERG 1972, DIAS 1982, KNOP et al. 1984)
coronafusene	(BALABAN 1982)

8.3 Anatomy

A pioneering work on some topological properties of coronoid systems is by POLANSKY and ROUVRAY (1977). A recent work of relevance to this topic is by HALL (1988).

In coronoid systems exactly the same modes of hexagons exist as in benzenoid systems (cf. Fig. 3.2).

An internal edge is shared by two hexagons and an internal vertex by three hexagons, in coronoids as well as in benzenoids. However, edges and vertices that are not internal are not called external in the case of coronoid systems. They may lie either on the outer or on the inner perimeter. Correspondingly we speak about (*outer* or *inner*) *boundary edges* and (*outer* or *inner*) *boundary vertices*.

Also, the structural features free edge, fissure, bay, cove and fjord (cf. Section 3.2 and especially Fig. 3.2) are recognized in coronoid systems. These features may occur either on the outer or on the inner perimeter.

As in the case of benzenoid systems, let h, n, m and n_i be used to designate the number of hexagons, vertices, edges, and internal vertices, respectively. These quantities are now (in coronoid systems) related via:

$$n = 4h - n_i,$$
$$m = 5h - n_i,$$
$$m = n + h.$$

Here again # will have the meaning "number of". Then

boundary vertices $= 4h - 2n_i$,
vertices of degree two $= 2h - n_i$,
vertices of degree three $= 2h$,
internal vertices of degree two $= 0$,
internal vertices of degree three $= n_i$,
boundary vertices of degree two $= 2h - n_i$,
boundary vertices of degree three $= 2h - n_i$,
internal edges $= h + n_i$,
boundary edges $= 4h - 2n_i$.

We wish to go into some detail with respect to the inner and outer perimeter. Hence, we need two more invariants. For this purpose let the corona hole be interpreted as a benzenoid system and let

h^0 = number of hexagons of the corona hole
n_i^0 = number of internal vertices of the corona hole

By virtue of the definition of a coronoid system one has $h^0 \geq 2$. Any benzenoid system with $h \geq 2$ may serve as a corona hole.
It was found that

outer boundary vertices $=$ # outer boundary edges
$= 4(h - h^0) - 2 - 2(n_i - n_i^0)$,
inner boundary vertices $=$ # inner boundary edges
$= 4h^0 + 2 - 2n_i^0$,
outer boundary vertices of degree two $= 2(h - h^0)$
$+ 2 - (n_i - n_i^0)$,
inner boundary vertices of degree two $= 2h^0 - 2 - n_i^0$,
outer boundary vertices of degree three
$= 2(h - h^0) - 4 - (n_i - n_i^0)$,
inner boundary vertices of degree three $= 2h^0 + 4 - n_i^0$.

The topological properties of primitive coronoid systems (see Section 8.2 for a definition) have been studied most extensively.
A primitive coronoid system consists of linearly and angularly annelated hexagons; they are in the modes L_2 and A_2, respectively. (L_2 modes may be missing.) The A_2 hexagons are also called *corners*. A corner may be *protruding* or *intruding*. A protruding corner has three edges on the outer perimeter, of which exactly one is a free edge. An intruding corner has three edges on the inner perimeter, one of them being a free edge. A primitive coronoid has at least six protruding corners. Let the number of protruding and intruding corners be denoted by J and I, respectively. Then

$$J = I + 6 ; \quad I = 0, 1, 2, \dots .$$

Below, we show a primitive coronoid with two intruding and eight protruding corners.

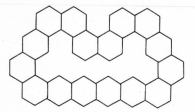

The number of corners is also the number of *segments*, say, S. Hence,

$$S = I + J = 2I + 6 .$$

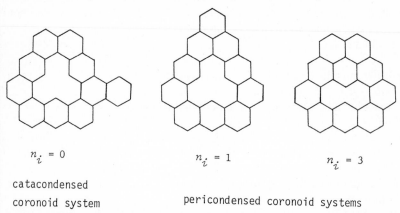

First classification: Catacondensed and pericondensed coronoid systems
    A catacondensed coronoid system has $n_i = 0$, while a pericondensed coronoid system has $n_i > 0$. This definition is equivalent to the one for catacondensed and pericondensed benzenoid systems; cf. Chapter 3.

$n_i = 0$

catacondensed
coronoid system

$n_i = 1$

$n_i = 3$

pericondensed coronoid systems

Again, as in benzenoid systems, if A_3-modes are absent, then the catacondensed system is unbranched. An unbranched catacondensed coronoid system is actually synonymous with a primitive coronoid system. The example below shows a branched catacondensed coronoid.

The coloring of vertices (cf. Section 3.3) is relevant also for coronoid systems. It is to be noted that peak(s) and valley(s) always occur on the outer perimeter. They may, but need not occur on the inner perimeter. With reference to Section 8.1, C_1 is seen to have one peak and one valley each on the inner perimeter, C_2 has none of either, while C_3 has one peak and no valleys.

The color excess Δ, defined as in benzenoid systems, is the absolute magnitude of the difference between the numbers of black and white vertices. It is also the absolute magnitude of the difference between the numbers of peaks and valleys.

Below we set up a schematic survey of different classification schemes for coronoid systems.

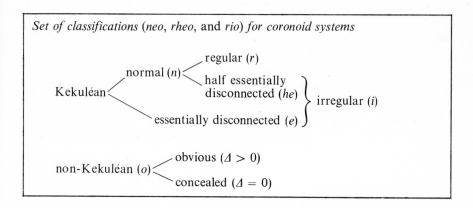

Set of classifications (neo, rheo, and *rio) for coronoid systems*

Here, the *neo* classification conforms with the corresponding concept for benzenoids (Chapter 3). The *rheo* classification refers to *regular* (*r*), *half essentially disconnected* (*he*), and non-Kekuléan (*o*) coronoid systems. The *rio* classification is coarser than the *rheo* classification. It refers to the categories regular (*r*), *irregular* (*i*), and non-Kekuléan (*o*).

Examples and definitions of the categories under the *rheo* and *rio* classifications are found in subsequent sections.

A coronoid system (like a benzenoid system) may belong to one of the symmetries $\mathbf{D}_{6h}$, $\mathbf{C}_{6h}$, $\mathbf{D}_{3h}$, $\mathbf{C}_{3h}$, $\mathbf{D}_{2h}$, $\mathbf{C}_{2h}$, $\mathbf{C}_{2v}$, and $\mathbf{C}_s$. The notions of central hexagon, central vertex, and central edge are also easily adapted to coronoid systems. It should also be noted that the center of a coronoid system is always situated in the corona hole. The subdivision of most of the symmetry groups as explained under "Third classification" for benzenoid systems (Section 3.7) is also applicable to coronoids.

The corona hole (interpreted as a benzenoid system) may have one of the eight symmetries mentioned above, but not necessarily the same as the coronoid system itself. If the symmetry group for the corona hole is $\mathbf{G}^0$, then the coronoid system has either the same symmetry ($\mathbf{G}^0$) or a lower symmetry in the sense of a subgroup of $\mathbf{G}^0$.

A segment is by definition a linear chain of hexagons between two neighboring corners inclusive. In other words, a corner belongs to two (neighboring) segments.

A primitive coronoid system with exactly six segments (or six corners) is called a *hollow hexagon* (BRUNVOLL, CYVIN and CYVIN 1987a). All the three systems C_1, C_2, and C_3 of Section 8.1 are hollow hexagons.

For every h value in a primitive coronoid system, there is a minimum and a maximum value for h^0, denoted by h^0_{min} and h^0_{max}, respectively. It was found (BRUNVOLL, CYVIN, et al. 1988a).

$$h^0_{min} = \begin{cases} \dfrac{1}{2}(h-4); & h = 8, 10, 12, \ldots \\[2mm] \dfrac{1}{2}(h-3); & h = 9, 11, 13, \ldots \end{cases}$$

and

$$\{12 \sqrt{h^0_{max} - 3}\} = h - 3 \,,$$

where the meaning of $\{x\}$ is the same as in Section 3.2. From the above relation it is found that

$$h^0_{max} = \left[\frac{1}{12}(h^2 - 6h + 12) \right],$$

where $[x]$ is used to denote the largest integer smaller than or equal to x. (For example, $[6.9] = 6$, $[7] = 7$, $[7.1] = 7$, $[\pi] = 3$.)

8.4 Enumeration

We start with reference to two papers which are devoted to the enumeration of coronoid systems in particular: KNOP et al. (1986b) and BRUNVOLL, CYVIN, and CYVIN (1987a). Special references to enumerations of primitive coronoids are found in a subsequent part of this section.

Tables 8.1–3 were produced from available enumeration data (BALABAN et al. 1987, CYVIN, CYVIN, and BRUNVOLL 1987, HE WJ et al. 1988). They may be compared with the first three tables of Chapter 4.

The principles of generating and enumerating coronoid systems are much the same as those for benzenoids (Chapter 4). The additions to primitive coronoid systems (cf. Section 8.2) have been exploited in particular (CYVIN, CYVIN, and BRUNVOLL 1987).

At this point some words should be said about the generation of regular coronoid systems, thus providing a definition of this category. In Section 4.5 the normal additions are defined: the added hexagon acquires the mode L_1, L_3, or L_5. Here we define the *regular additions* as the normal additions plus the corona-condensations (cf. Section 8.2). The corona-condensations are of two types, linear and angular, pertaining to the modes L_2 and A_2, respectively. Assume now a generation procedure of benzenoid and coronoid systems implying successively all possible regular additions. In this way all normal benzenoid systems are generated (according to a conjecture) in addition to a selection of Kekuléan coronoid systems, which by definition are the regular coronoids. Thus not all normal (Kekuléan) coronoid systems can be generated by regular addition. Those Kekuléan coronoid systems which are not regular are said to be irregular.

Table 8.1. Numbers of coronoid systems belonging to the different symmetry groups

h	D_{6h}	C_{6h}	D_{3h}	C_{3h}	D_{2h}	C_{2h}	C_{2v}	C_s
8	0	0	0	0	1	0	0	0
9	0	0	1	0	0	0	1	3
10	0	0	0	0	3	3	9	28
11	0	0	0	0	0	0	24	259
12	1	0	3	2	10	25	79	1834
13	0	0	0	0	0	0	185	12178
14	0	0	0	0	*	*	*	*
15	0	0	9	17	0	0	*	*
16	0	0	0	0	*	*	*	*
17	0	0	0	0	0	0	*	*
18	3	1	23	102	*	*	*	*

* Unknown

Table 8.2. Numbers of coronoid systems according to the first classification

h	Catacondensed	Pericondensed	Total
8	1	0	1
9	3	2	5
10	15	28	43
11	62	221	283
12	312	1642	1954
13	1435	10928	12363

Table 8.3. Numbers of coronoid systems according to the *rheo* and *rio* classifications*

h	r	he	e	i	o
8	1	0	0	0	0
9	3	0	0	0	2
10	18	6	0	6	19
11	90	36	2	38	155
12	526	289	39	328	1100
13	2810	**	**	2240	7313

* e = essentially disconnected; he = half essentially disconnected; i = irregular; o = non-Kekuléan; r = regular (i = he + e)
** Unknown

The definition of essentially disconnected coronoid systems is clear. These systems have fixed double and/or single bonds. Now it may serve as a definition of half essentially disconnected coronoid systems (CYVIN, CYVIN, and BRUNVOLL 1987) to state that these systems are irregular and not essentially disconnected. The reason for the name "half essentially disconnected" is explained in the subsequent section on KEKULÉ structures.

Table 8.4. Numbers of primitive coronoid systems classified according to symmetry

h	D_{6h}	C_{6h}	D_{3h}	C_{3h}	D_{2h}	C_{2h}	C_{2v}	C_s	Total primitive
8	0	0	0	0	1	0	0	0	1
9	0	0	1	0	0	0	0	0	1
10	0	0	0	0	2	0	1	0	3
11	0	0	0	0	0	0	1	1	2
12	1	0	2	0	2	2	2	2	11
13	0	0	0	0	0	0	4	8	12
14	0	0	0	0	6	5	12	17	40
15	0	0	2	2	0	0	6	58	68
16	0	0	0	0	7	18	38	129	192
17	0	0	0	0	0	0	23	372	395
18	2	0	5	4	15	48	92	895	1061
19	0	0	0	0	0	0	55	2377	2432
20	0	0	0	0	20	137	272	5889	6318
21	0	0	5	15	0	0	142	15255	15417
22	0	0	0	0	46	363	705	38546	39660
23	0	0	0	0	0	0	367	99183	99550
24	2	1	17	34	50	992	1872	253420	256388
25	0	0	0	0	0	0	973	653030	654003

Table 8.5. Numbers of primitive coronoid systems with hexagonal and trigonal symmetries for $27 \le h \le 66$ (in continuation of Table 8.4)

h	D_{6h}	C_{6h}	D_{3h}	C_{3h}
27	0	0	13	104
30	5	2	44	258
33	0	0	35	*
36	5	8	127	*
39	0	0	92	*
42	12	19	339	*
45	0	0	246	*
48	13	55	929	*
51	0	0	658	*
54	31	138	2508	*
57	0	0	1778	*
60	33	373	6839	*
63	0	0	4805	*
66	80	957	*	*

* Unknown

Table 8.3 gives the numbers of coronoid systems in the frames of the *rheo* classification for $h \leq 12$. For $h = 13$ the numbers of half essentially disconnected and essentially disconnected coronoids are not known separately, but their sum is known and gives the number of irregular systems; hence, the *rio* classification is established.

The class of primitive coronoid systems has been enumerated most extensively (CYVIN, CYVIN, BRUNVOLL, and BERGAN 1987, BALABAN et al. 1987, CYVIN, BRUNVOLL, and CYVIN 1988c, 1989, BRUNVOLL, CYVIN et al. 1988a, CYVIN 1989). Table 8.4 shows the numbers for $h \leq 25$. The distribution of the systems into symmetry groups is included. It is also possible (as for benzenoid systems) to generate and enumerate coronoid systems specifically for different symmetries. Table 8.5 shows the results of such enumerations for the hexagonal and trigonal systems.

8.5 KEKULÉ **Structures**

Some work has been done on the enumeration of KEKULÉ structures for coronoid systems. Most of it is based on the method of fragmentation (Paragraph 5.1.1), which is applicable to coronoid as well as benzenoid systems. We have seen in the cited paragraph how the method leads to fragments with edges not belonging to hexagons. In the example below we find such edges as part of a macrocycle.

where

Here, we have not found it necessary to carry out all the intermediate steps because the principles are the same as in benzenoid systems (Paragraph 5.1.1). We started from ovalene, of which the KEKULÉ structure count K, is known to be 50. It is obtained, for instance, from formula (5.8) by realizing that ovalene is $O(3, 2, 2)$. Hence, the primitive coronoid among the first fragments, viz. cyclo[*d.e.e.e.d.e.e.e*]-octakisbenzene, has $K = 50 - 10 = 40$.

Consider, in analogy with the zigzag benzenoid chain, h angularly annelated hexagons in a macrocyclic arrangement. Denote this system by $a(h)$. These systems have been studied by BERGAN et al. (1987). Coronoid systems of the type $a(h)$ can be constructed with $h = 12, 14, 16, \ldots$. Two examples are:

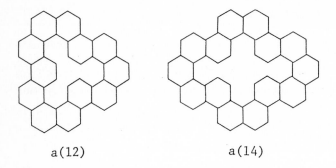

a(12)   a(14)

One has the formula for KEKULÉ structure counts:

$$K\{a(h)\} = F_h + F_{h-2} + 2 ,$$

where F_k are the FIBONACCI numbers already encountered in Paragraph 5.1.2. The explicit formula reads:

$$K\{a(2p)\} = \left(\frac{3 + \sqrt{5}}{2}\right)^p + \left(\frac{3 - \sqrt{5}}{2}\right)^p + 2 , \tag{8.1}$$

where $h = 2p$. This formula is consistent with the recurrence relation

$$K\{a(2p)\} = 3K\{a(2p - 2)\} - K\{a(2p - 4)\} - 2 . \tag{8.2}$$

The class $a(h)$ has been generalized (BERGAN et al. 1987) to $w(j, N)$, consisting of N equidistant segments of length j (in terms of the number of hexagons), again in a macrocyclic arrangement. In this notation, $a(h) = w(2, h)$. Coronoid systems can be constructed for $N = 6, 10, 12, 14, 16 \ldots$ when $j > 2$. Such a system has N corners, and the number of hexagons is $h = (j - 1) N$. Of special interest is the case with $j = 3$, since kekulene belongs to this class; specifically $w(3, 6) = $ kekulene. For this case it was found that

$$K\{w(3, 2p)\} = (3 + \sqrt{8})^p + (3 - \sqrt{8})^p + 2 . \tag{8.3}$$

On inserting $p = 3$ in (8.3), one indeed obtains $K\{kekulene\} = 200$.

Kekulene is also a member of the class $w(j, 6)$. For $j > 2$ these systems are recognized as the hollow hexagons with equidistant segments, while $j = 2$ corresponds to coronene, a pericondensed benzenoid system:

$K = 20$

$K = 200$

$K = 1300$

For this class, the combinatorial K formula was derived independently in three works, all published the same year (BERGAN et al. 1986, BABIĆ and GRAOVAC 1986, HOSOYA 1986). The result is

$$K\{w(a + 1, 6)\} = (a^3 + 3a)^2 + 4 = (a^2 + 1)^2 (a^2 + 4)$$
$$= a^6 + 6a^4 + 9a^2 + 4 . \tag{8.4}$$

Here, $j = a + 1$. One obtains, of course, kekulene for $a = 2$ ($K = 200$) and coronene ($K = 20$) for $a = 1$.

The class $w(j, 6)$ considered above is a special case of $o(p, q, r)$, the hollow hexagons with pairwise parallel segments, as shown below.

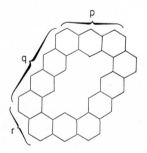

p

q

r

$$K\{o(p, q, r)\} = K\{o(a + 1, b + 1, c + 1)\} = (abc + a + b + c)^2 + 4. \tag{8.5}$$

The given formula (8.5) is due to BERGAN et al. (1986).

The two combinatorial formulas (8.3) and (8.4) pertain to $w(j, N)$ with a fixed value of j ($j = 3$) and with a fixed value of N ($N = 6$), respectively. This reminds one of the situation for oblate rectangles (described in Paragraph 5.1.3). However, in the present case we are better off, inasmuch as a general combinatorial formula for $K\{w(j, N)\}$ is available (BERGAN et al. 1987):

$$K\{w(a + 1, N)\} = \left(\frac{a + \sqrt{a^2 + 4}}{2}\right)^N + \left(\frac{a - \sqrt{a^2 + 4}}{2}\right)^N + 2 \qquad (8.6)$$

The following procedure is applicable to primitive coronoid systems in general (BERGAN et al. 1986):

(a) Delete an arbitrary corner (either protruding or intruding). Let k denote the KEKULÉ structure count for the remaining unbranched catacondensed benzenoid; it can be found, for instance, by the numeral-in-hexagon algorithm (Paragraph 5.1.4).

(b) Delete the same corner with its two neighboring segments. Let k' denote the KEKULÉ structure count of the remaining benzenoid, to which again the numeral-in-hexagon algorithm is applicable.

(c) The total KEKULÉ structure count is

$$K = k + k' + 2$$

We illustrate this algorithm for a primitive coronoid system with $K = 584$.

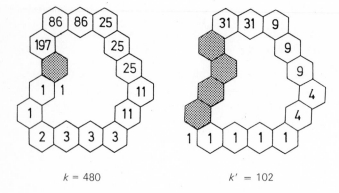

$k = 480$ $k' = 102$

The two extra KEKULÉ structures, which do not belong to $k + k'$, are *annulenoid*; here all internal edges correspond to single bonds, and there are no aromatic sextets present:

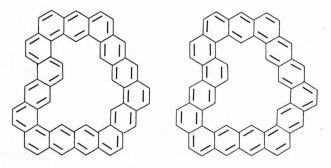

It should be noted that there exist two more Kekulé structures with single bonds at all internal edges:

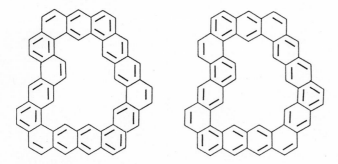

These two systems do possess aromatic sextets. The left- and right-hand side systems depicted above are contained in the k and k' Kekulé structures, respectively.

So far we have treated the Kekulé structures only for primitive coronoid systems, which are catacondensed. Below, we give a combinatorial K formula for a class of pericondensed coronoid systems, where each member has exactly two internal vertices.

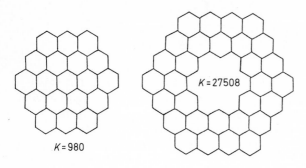

$$K = 2(1 + pq) \qquad (8.7)$$

A more advanced example is the class of circumkekulene and its homologs (Cyvin, Cyvin, and Brunvoll 1988):

$K = 980$

$K = 27508$

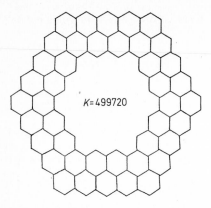

Let a be defined so that the number of hexagons in the top (or bottom) row is $a + 2$. Then the K formula reads:

$$K = \frac{1}{32}(a^2 + 4)(a^2 + 2a + 5)(a^8 + 4a^7 + 22a^6$$
$$+ 52a^5 + 129a^4 + 176a^3 + 208a^2 + 128a + 64) \tag{8.8}$$

Below, we show two examples of essentially disconnected coronoid systems. The K numbers are given as products of the KEKULÉ structure counts for the effective units. Notice that one of the effective units in the left-hand side example has edges not belonging to hexagons, which constitute a part of a macrocycle.

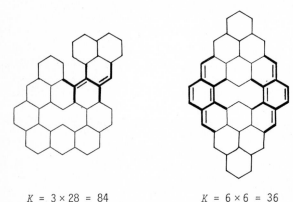

$$K = 3 \times 28 = 84 \qquad\qquad K = 6 \times 6 = 36$$

Finally in this section it remains to show some examples of half essentially disconnected coronoid systems. This concept was first encountered under a set of classifications in Section 8.3, while a definition was provided in Section 8.4. In a half essentially disconnected coronoid system the KEKULÉ structures may be divided into two sets, each possessing fixed bonds. The two sets together should

It should be noted that there exist two more KEKULÉ structures with single bonds at all internal edges:

These two systems do possess aromatic sextets. The left- and right-hand side systems depicted above are contained in the k and k' KEKULÉ structures, respectively.

So far we have treated the KEKULÉ structures only for primitive coronoid systems, which are catacondensed. Below, we give a combinatorial K formula for a class of pericondensed coronoid systems, where each member has exactly two internal vertices.

$$K = 2(1 + pq) \qquad\qquad (8.7)$$

A more advanced example is the class of circumkekulene and its homologs (CYVIN, CYVIN, and BRUNVOLL 1988):

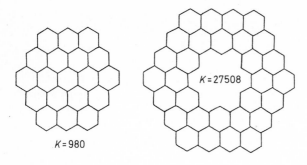

$K = 980$

$K = 27508$

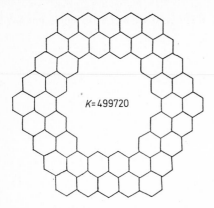

Let a be defined so that the number of hexagons in the top (or bottom) row is $a + 2$. Then the K formula reads:

$$K = \frac{1}{32}(a^2 + 4)(a^2 + 2a + 5)(a^8 + 4a^7 + 22a^6$$
$$+ 52a^5 + 129a^4 + 176a^3 + 208a^2 + 128a + 64) \tag{8.8}$$

Below, we show two examples of essentially disconnected coronoid systems. The K numbers are given as products of the KEKULÉ structure counts for the effective units. Notice that one of the effective units in the left-hand side example has edges not belonging to hexagons, which constitute a part of a macrocycle.

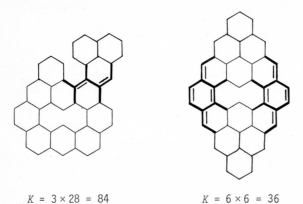

$$K = 3 \times 28 = 84 \qquad\qquad K = 6 \times 6 = 36$$

Finally in this section it remains to show some examples of half essentially disconnected coronoid systems. This concept was first encountered under a set of classifications in Section 8.3, while a definition was provided in Section 8.4. In a half essentially disconnected coronoid system the KEKULÉ structures may be divided into two sets, each possessing fixed bonds. The two sets together should account for the complete set of KEKULÉ structures. The examples below are supposed to elucidate this characterization.

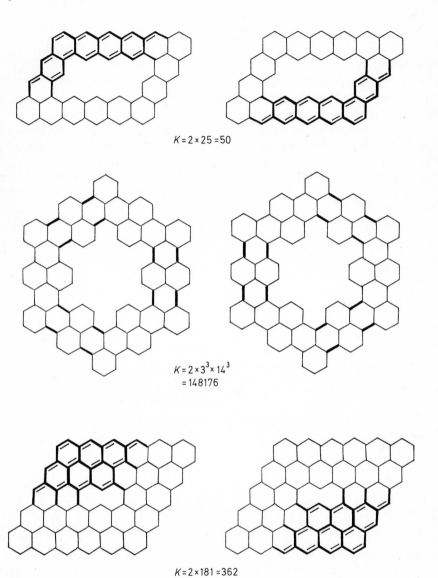

$$K = 2 \times 25 = 50$$

$$K = 2 \times 3^3 \times 14^3$$
$$= 148176$$

$$K = 2 \times 181 = 362$$

The first example (top row) is a member of the class pertaining to (8.7). All members of this class are half essentially disconnected coronoid systems. On this ground it is easy to derive the given K formula (8.7).

It has not been proved that the above characterization in terms of two schemes for KEKULÉ structures is sound for half essentially disconnected coronoid systems in general. However, no counterexamples have been detected, in spite of extensive computer-aided generations and classifications of coronoid systems.

8.6 CLAR **Structures and All-Coronoids**

The definitions of CLAR structures (Paragraph 7.2.1) and the equivalent to all-benzenoid systems (Paragraph 7.2.3) are also applicable to coronoid systems. The analogues to all-benzenoid systems are called *all-coronoid systems*. In the following we only give a few examples of these concepts.

For the two cycloarenes which have been synthesized, C_1 and C_2 (see Section 8.1), the numbers of aromatic sextets in the CLAR structures are 6 and 4, respectively. There is a marked difference between the two systems C_1 and C_2. In the former case the CLAR structure is unique:

For C_2, on the other hand, there are nine CLAR structures. They are specified below, where only those not being symmetrically equivalent are depicted. The numbers of symmetrically equivalent structures are indicated by the inscribed numerals.

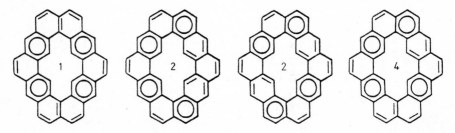

The sextet patterns of cyclo[*d.e.e.e.d.e.e.e.*]octakisbenzene have been considered in detail (OHKAMI et al. 1981). This system shares that property with kekulene of having a unique CLAR structure, which, however, possesses 4 aromatic sextets:

The above system, as well as kekulene, is easily converted into an all-coronoid system by annelations of hexagons to the free edges of the corners. These two all-coronoid systems (BERGAN et al. 1987, CYVIN, BERGAN, and CYVIN 1987) are shown below along with a third example.

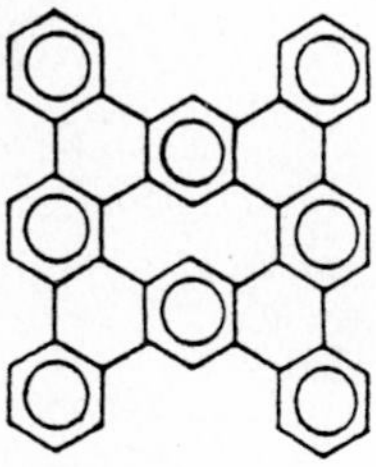

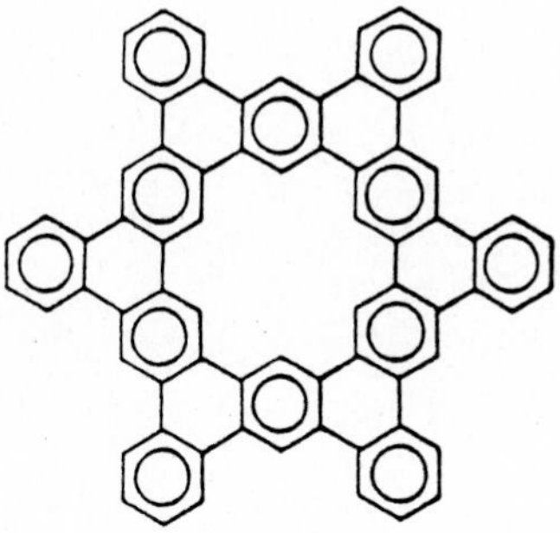

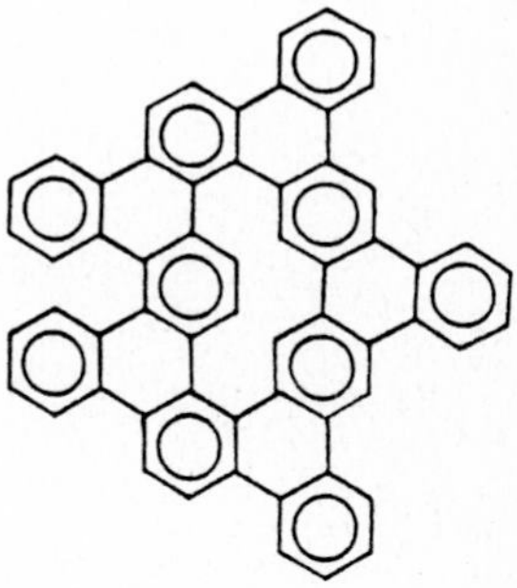

These three systems are all catacondensed. Naturally, pericondensed all-coronoids also can be constructed in many ways. Three examples are depicted below.

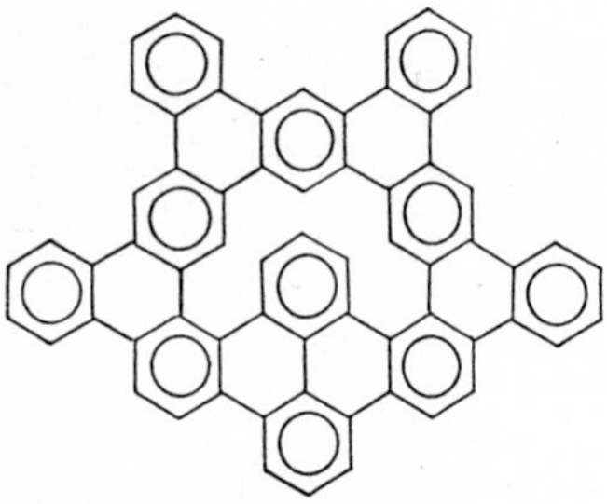

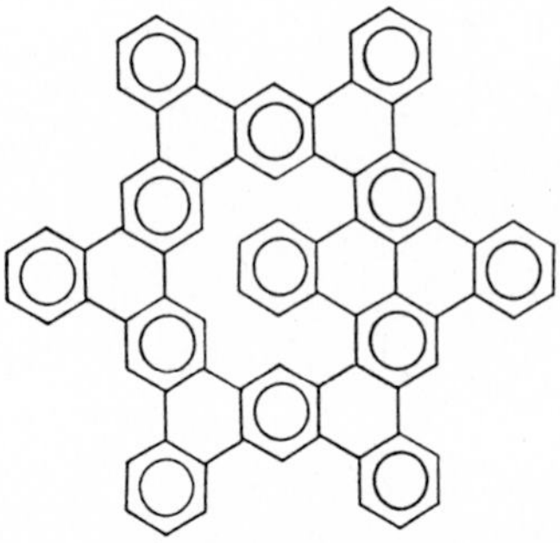

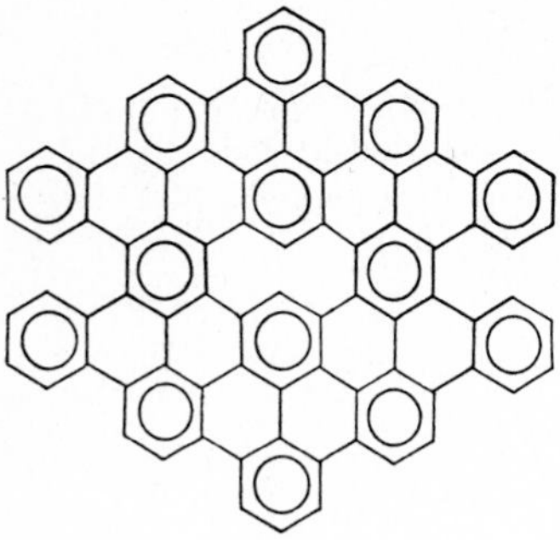

Reading Section 8.6 one may observe that it contains only definitions and examples. This is because at present no general result involving CLAR type structures is known for coronoid systems.

In general, the development of chemical theories or the discovery of mathematical results which would point out some of the distinguishing (chemical or mathematical) properties of coronoids remains a challenge for the future.

Chapter 9

Back to Chemistry

9.1 Thermodynamic Properties of Benzenoid Hydrocarbons

There is no doubt that the knowledge of the thermodynamic data for benzenoid hydrocarbons is of prime importance for the understanding of their basic physical and chemical properties and, in particular, their formation in high-temperature processes. Furthermore, thermodynamic parameters are needed for testing the great many resonance energies which are the outcomes of various theoretical approaches to benzenoid molecules (some of which are outlined in Chapters 5, 6, and 7).

Relatively little is known about the thermodynamic properties of benzenoid hydrocarbons.* This is mainly the consequence of the lack of sufficiently pure samples of these compounds. Even in such an easy case as anthracene, the reported values for the gas phase enthalpy of formation at 298 K range from 209 kJ mol^{-1} to 234 kJ mol^{-1}. These discrepancies can be explained by different purities of anthracene samples used in the measurements (SOMAYAJULU and ZWOLINSKI 1974). As shown with the example of dicoronylene, $C_{48}H_{20}$, the purification of benzenoids with larger molecular masses is very difficult (LEMPKA et al. 1985).

Bearing the above in mind, it is not surprising that various procedures for the calculation of the thermodynamic properties of benzenoid hydrocarbons were developed (SOMAYAJULU and ZWOLINSKI 1974, STEIN, GOLDEN, and BENSON 1977, and the references cited therein).

Here, we present the both very simple and quite accurate method of STEIN, GOLDEN, and BENSON (1977). This method requires four empirical parameters and is designed to reproduce both the standard gas-phase enthalpies of formation (ΔH^0_{f298}), entropies (S^0_{298}), and heat capacities ($C_{p^0_T}$) at various temperatures.

Each of the above mentioned quantities is presented in the form

$$t_1 X_1 + t_2 X_2 + t_3 X_3 + t_4 X_4 \, , \tag{9.1}$$

where X_1, X_2, X_3, X_4 are empirical parameters whose values are given in Table 9.1.

---

* For instance, the many-volume TRC Thermodynamic Tables contain data for only 24 benzenoid hydrocarbons.

The multipliers t_1, t_2, t_3, t_4 depend on the structure of the benzenoid hydrocarbon considered. Using the terminology of Chapter 3,

$t_1 = \#$ vertices of degree two,
$t_2 = \#$ external vertices of degree three lying in a fissure $= \#$ fissures,
$t_3 = \#$ external vertices of degree three lying in a bay $= 2 \times \#$ bays,
$t_4 = \#$ internal vertices.

The method of STEIN, GOLDEN, and BENSON (1977) does not allow for benzenoids with coves and fjords. Because each fissure contains exactly one vertex of degree three, t_2 is simply equal to the number of fissures. For similar reasons t_3 is twice the number of bays.

Table 9.1. The empirical parameters used in the model of STEIN, GOLDEN, and BENSON (1977); for explanation see text

	$\Delta H^0_{f\,298}$ kJ mol^{-1}	S^0_{298} JK^{-1} mol^{-1}	$C_{p\,T}^0$ J K^{-1} mol^{-1}					
			300 K	400 K	500 K	600 K	800 K	1000 K
X_1	13.81	48.24	13.56	18.58	22.84	26.36	31.55	35.19
X_2	20.08	−20.92	12.51	15.31	17.66	19.41	21.88	23.22
X_3	15.48	−20.92	12.51	15.31	17.66	19.41	21.88	23.22
X_4	6.07	7.61	8.70	14.64	14.64	16.86	19.87	21.51

From Table 9.1 it can be seen that for the entropy and heat capacities, $X_2 = X_3$. This means that in these cases, (9.1) reduces to a three-parametric expression:

$$t_1 X_1 + (t_2 + t_3) X_2 + t_4 X_4 ,$$

where

$$t_2 + t_3 = \# \text{ external vertices of degree three.}$$

Using formulas of type (9.1) it was possible to estimate the equilibrium concentrations of benzenoid hydrocarbons in the presence of acetylene and hydrogen over the range 1400–3000 K (STEIN 1978).

9.2 More on Non-Kekuléan Benzenoid Hydrocarbons

In Chapter 5 we described in detail the unsuccessful attempts of CLAR to prepare triangulene (I) and mentioned that neither of his efforts to synthesize dibenzo[de, hi]-naphthacene (II) and dibenzo[de, jk]pentacene (III) gave any results.

I   II   III

In spite of the great progress that synthetic organic chemistry has made in the last two decades, the non-Kekuléan hydrocarbons I–III are still out of its range.

In 1977 ICHIRO MURATA with coworkers made an important step forward in this direction by synthesizing the dianion of I as well as the dication and the dianion of III (HARA, TANAKA, et al. 1977, HARA, YAMAMOTO, and MURATA 1977).

Triangulene-4,8-quinone (IV) was reduced with LiAlH$_4$–AlCl$_3$ complex in ether at room temperature to give 3,8-dihydrotriangulene (V), which is a stable compound. The solution of V in tetrahydrofuran was treated with n-buthyl-lithium at —78 °C under nitrogen. The dianion (VI) of triangulene was obtained (HARA, TANAKA, et al. 1977).

The structure of VI was inferred from the NMR spectrum and from the fact that addition of D$_2$O produced the doubly deuterated 3,8-dihydrotriangulene (VII). The NMR spectrum shows that the triangulene dianion has D_{3h} symmetry.

The dianion (X) of dibenzo[de, jk]pentacene was obtained using a completely analogous method, starting with dibenzo[de, jk]pentacene-7,9-quinone (VIII) (HARA, YAMAMOTO, and MURATA 1977):

The dication (XI) of dibenzo[*de, jk*]pentacene was obtained by dissolving IX in concentrated sulfuric acid. A dark green solution resulted, whose NMR spectrum showed the existence of XI; the NMR data were compatible with the C_{2v} symmetry of the dication XI (HARA, YAMAMOTO, and MURATA 1977).

When the hydrocarbon IX was reacted with trityl tetrafluoroborate, a black solid was obtained, believed to contain the monocation XII.

In any case, when the black substance dissolved in concentrated sulfuric acid, the solution was shown to contain the dication XI (HARA, YAMAMOTO, and MURATA 1977).

9.3 Extraterrestrial Benzenoid Hydrocarbons

From the data presented in Section 1.3 it can be concluded that benzenoid hydrocarbons are present practically everywhere on the surface of the Earth. In the same section it was pointed out that benzenoid hydrocarbons were found also in meteorites, whose extraterrestrial origin is evident.

For some time, astrophysicists have studied the infrared radiation coming from the Universe. Their main finding was a series of discrete emission bands, apparently having their origin in the interstellar clouds. These bands are always observed as a set, and the most intense maxima are about 3.3, 6.2, 7.7, 8.6 and 11.3 μm.

In 1981 it was recognized that precisely the same wavelengths are characteristic of the vibrational spectra of large polycyclic aromatic hydrocarbons, in particular, large benzenoids. Eventually it was generally accepted that the infrared radiation is caused by PAHs, located in interstellar clouds (LÉGER et al. 1987). The actual structure of the compounds (or perhaps compound) responsible for the radiation remained, however, obscure.

ALLAMANDOLA et al. (1985) pointed out the close resemblance between the astronomical infrared spectrum and the RAMAN spectrum of automobile exhaust, which is known to consist of a complex mixture of PAHs and particles of elemental carbon. According to ALLAMANDOLA et al. (1985), the infrared emission spectrum is best explained as coming from a collection of PAHs, and not from an individual compound.

Nevertheless, efforts have been made to find out the structure of the compounds which are the main constituents of the interstellar PAH mixture.

It is clear that the physical conditions in the interstellar clouds are harsh and that only the most stable PAHs have chance to survive. Owing to intense UV radiation, the hydrocarbon molecules are probably in an electronically excited and/or ionized state.

The all-benzenoid hydrocarbons are both thermodynamically and photochemically the most stable benzenoids (see Chapter 7). Bearing this in mind, SCHMIDT proposed hexabenzo[bc, ef, hi, kl, no, qr]coronene (XIII) and tribenzo[a, g, m]coronene (XIV) as suitable candidates for extraterrestrial benzenoid hydrocarbons (HENDEL et al. 1986, OBENLAND and SCHMIDT 1987).

XIII   XIV

AIHARA (1987) challenged SCHMIDT's suggestion by pointing out that the 11.3 μm band, which is assigned to the out-of-plane C–H bending vibrations, is very sensitive to the number of hydrogen atoms attached to the respective six-membered ring of the benzenoid hydrocarbon. The wavelength and the intensity of this band indicate that the majority of the hydrogen atoms are solo, the amount of duo and trio hydrogens is much smaller, whereas quartet hydrogens are absent.*

Solo   Duo   Trio   Quartet

---

* Using the terminology of Chapter 3, it is seen that solo hydrogens require P_4 mode hexagons, duo and trio hydrogens require hexagons of mode L_3 and P_2, respectively, while quartet hydrogens occur on hexagons of mode L_1. Thus the benzenoid systems required by AIHARA's analysis of the 11.3 μm band must be composed of L_6 and P_4 mode hexagons and a few (but at least six) hexagons of L_3 mode.

From this AIHARA concluded that neither XIII nor XIV (having not one solo hydrogen) are present in the interstellar clouds. His calculations pointed to circumnaphthacene (XV) and circumanthracene (XVI) as the benzenoid hydrocarbons that best satisfy the requirement of stable ground, excited and ionized states, and whose vibrational spectra match with the observed infrared emission data (AIHARA 1987).

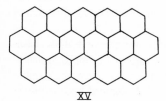

XV

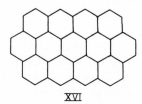

XVI

The opinion of the authors of this book is that a definite answer to this question is not to be expected soon. On the other hand, one should not exclude the possibility that the interstellar PAHs are mixtures of hundreds or thousands of polycyclic benzenoid and nonbenzenoid hydrocarbons, none of which are present in significant excess.

It has been estimated (ALLAMANDOLA et al. 1985) that the abundance of polycyclic aromatic hydrocarbons in the interstellar clouds is 2×10^{-7} times the abundance of hydrogen. From this follows that 1–2% of all of the available carbon is in the form of benzenoid hydrocarbons. Bearing in mind the enormous size of the interstellar clouds, we are tempted to conclude that benzenoid hydrocarbons are among the most frequently occurring organic molecules in the Universe.

References

Aboav D, Gutman I (1988) Chem. Phys. Letters 148: 90
Aida M, Hosoya H (1980) Tetrahedron 36: 1321
Aihara J (1977) Bull. Chem. Soc. Japan 50: 2010
Aihara J (1987) Bull. Chem. Soc. Japan 60: 3143
Allamandola LJ, Tielens AGGM, Barker JR (1985) Astrophys. J. 290: L25
Armit JW, Robinson R (1925) J. Chem. Soc.: 1604
Babić D, Graovac A (1986) Croat. Chem. Acta 59: 731
Badger GM (1969) Aromatic Character and Aromaticity. Cambridge Univ. Press, Cambridge
Balaban AT (1982) Pure Appl. Chem. 54: 1075
Balaban AT, Brunvoll J, Cioslowski J, Cyvin BN, Cyvin SJ, Gutman I, He WC, He WJ, Knop JV, Kovačević M, Müller WR, Szymanski K, Tošić R, Trinajstić N (1987) Z. Naturforsch. 42a: 863
Balaban AT, Brunvoll J, Cyvin BN, Cyvin SJ (1988) Tetrahedron 44: 221
Balaban AT, Harary F (1968) Tetrahedron 24: 2505
Balasubramanian K, Kaufman JJ, Koski WS, Balaban AT (1980) J. Comput. Chem. 1: 149
Bergan JL, Cyvin BN, Cyvin SJ (1987) Acta Chim. Hung. 124: 299
Bergan JL, Cyvin SJ, Cyvin BN (1986) Chem. Phys. Letters 125: 218
Biermann D, Schmidt W (1980) Israel J. Chem. 20: 312
Binsch G (1973) Naturwiss. 60: 369
Bjørseth A, Ramdahl T (1985) in: Bjørseth A, Ramdahl T (eds) Handbook of Polycyclic Aromatic Hydrocarbons. Marcel Decker, New York, p 1
Blouke MM, Cowens MW, Hall JE, Westphal JA, Christensen AB (1980) Appl. Opt. 19: 3318
Blumer M (1975) Chem. Geol. 16: 247
Blumer M (1976) Sci. Amer. 234: 35
Bonchev D, Balaban AT (1981) J. Chem. Inf. Comput. Sci. 21: 223
Bräuchle C, Kabza H, Voitländer J (1980) Chem. Phys. 48: 369
Bräuchle C, Voitländer J (1982) Tetrahedron 38: 285
Brown RL (1983) J. Comput. Chem. 4: 556
Brunvoll J, Cyvin BN, Cyvin SJ (1987a) J. Chem. Inf. Comput. Sci. 27: 14
Brunvoll J, Cyvin BN, Cyvin SJ (1987b) J. Chem. Inf. Comput. Sci. 27: 171
Brunvoll J, Cyvin BN, Cyvin SJ, Gutman I (1988a) Match 23: 209
Brunvoll J, Cyvin BN, Cyvin SJ, Gutman I (1988b) Z. Naturforsch. 43a: 889
Brunvoll J, Cyvin BN, Cyvin SJ, Gutman I, Tošić R, Kovačević M (1989) J. Mol. Struct. (Theochem) 184: 165
Brunvoll J, Cyvin SJ, Cyvin BN, Gutman I, He WJ, He WC (1987) Match 22: 105
Chandra H, Symons MCR, Griffiths DR (1988) Nature 332: 526
Cioslowski J (1986) Match 20: 95
Cioslowski J (1988) to be published; quoted by Aboav and Gutman (1988)
Cioslowski J, Polansky OE (1988) Theor. Chim. Acta 74: 55
Cioslowski J, Wala M (1986) Match 21: 195
Clar E (1964a) *Polycyclic Hydrocarbons*. Academic Press, London
Clar E (1964b) Chimia 12: 375

Clar E (1972) *The Aromatic Sextet.* Wiley, London
Clar E, Ironside CT, Zander M (1959) J. Chem. Soc.: 142
Clar E, Kemp W, Stewart DG (1958) Tetrahedron 3: 325
Clar E, Stewart DG (1953) J. Am. Chem. Soc. 75: 2667
Clar E, Zander M (1958) J. Chem. Soc.: 1861
Collin G, Zander M (1983) Chemie uns. Zeit 17: 181
Cook JW, Hewett CL, Hieger I (1933) J. Chem. Soc.: 395
Cooper DL, Gerratt J, Raimondi M (1986) Nature 323: 699
Cvetković D, Doob M, Gutman I, Torgašev A (1988) *Recent Results in the Theory of Graph Spectra.* North-Holland, Amsterdam
Cvetković D, Gutman I, Trinajstić N (1974) J. Chem. Phys. 61: 2700
Cyvin BN, Brunvoll J, Cyvin SJ, Gutman I (1988) Match 23: 163
Cyvin BN, Cyvin SJ, Brunvoll J (1988) Monatsh. Chem. 119: 563
Cyvin SJ (1989) Monatsh. Chem. 120: 243
Cyvin SJ, Bergan JL, Cyvin BN (1987) Acta Chim. Hung. 124: 691
Cyvin SJ, Brunvoll J, Cyvin BN (1988a) J. Mol. Struct. (Theochem) 180: 329
Cyvin SJ, Brunvoll J, Cyvin BN (1988b) Match 23: 189
Cyvin SJ, Brunvoll J, Cyvin BN (1988c) Acta Chem. Scand. A42: 434
Cyvin SJ, Brunvoll J, Cyvin BN (1989) Comp. & Math. with Appl. 17: 355
Cyvin SJ, Cyvin BN, Brunvoll J (1987) Chem. Phys. Letters 140: 124
Cyvin SJ, Cyvin BN, Brunvoll J, Bergan JL (1987) Coll. Sci. Papers Fac. Sci. Kragujevac 8: 137
Cyvin SJ, Gutman I (1986a) Z. Naturforsch. 41a: 1079
Cyvin SJ, Gutman I (1986b) Match 19: 229
Cyvin SJ, Gutman I (1987) J. Mol. Struct. (Theochem) 150: 157
Cyvin SJ, Gutman I (1988) *Kekulé Structures in Benzenoid Hydrocarbons.* Springer, Berlin Heidelberg New York
Dewar MJS, de Llano C (1969) J. Am. Chem. Soc. 91: 789
Dewar MJS, Longuet-Higgins HC (1952) Proc. Roy. Soc. London A214: 482
Dias JR (1982) J. Chem. Inf. Comput. Sci. 22: 15
Dias JR (1987) *Handbook of Polycyclic Hydrocarbons. Part A. Benzenoid Hydrocarbons.* Elsevier, Amsterdam
Diederich F, Staab HA (1978) Angew. Chem. Int. Ed. 17: 372
Dopper JE, Wynberg H (1972) Tetrahedron Letters: 763
Eilfeld P, Schmidt W (1981) J. Electron Spectr. Rel. Phenom. 24: 101
El-Basil S (1984a) Theor. Chim. Acta 65: 191
El-Basil S (1984b) Theor. Chim. Acta 65: 199
El-Basil S (1986a) Theor. Chim. Acta 70: 53
El-Basil S (1986b) J. Chem. Soc. Faraday II 82: 299
El-Basil S (1987) J. Math. Chem. 1: 153
El-Basil S (1988) Discrete Appl. Math. 19: 145
El-Basil S, Randić M (1987) J. Math. Chem. 1: 281
Franck HG, Stadelhofer JW (1987) *Industrielle Aromatenchemie. Rohstoffe, Verfahren, Produkte.* Springer, Berlin Heidelberg New York
Funhoff DJH, Staab HA (1986) Angew. Chem. Int. Ed. 25: 742
Geissman TA, Sim KY, Murdoch J (1967) Experientia 23: 793
George P (1975) Chem. Rev. 75: 85
Gerratt J (1987) Chem. Britain 23: 327
Golomb SW (1954) Am. Math. Monthly 61: 675
Golomb SW (1965) *Polyominoes.* Scribner, New York
Gomes JANF (1980) Croat. Chem. Acta 53: 561
Gomes JANF (1981) Theor. Chim. Acta 59: 333
Gordon M, Davison WHT (1952) J. Chem. Phys. 20: 428
Graovac A, Gutman I, Trinajstić N, Živković T (1972) Theor. Chim. Acta 26: 67
Gutman I (1974) Croat. Chem. Acta 46: 209
Gutman I (1977a) Match 3: 121

Gutman I (1977b) Theor. Chim. Acta 45: 309
Gutman I (1978) Bull. Chem. Soc. Japan 51: 2729
Gutman I (1981) Match 11: 127
Gutman I (1982a) Bull. Soc. Chim. Beograd 47: 453
Gutman I (1982b) Z. Naturforsch. 37a: 69
Gutman I (1985) J. Serb. Chem. Soc. 50: 451
Gutman I (1986) Match 21: 317
Gutman I (1990) Z. phys. Chem. (Leipzig) in press
Gutman I, Bosanac S (1977) Tetrahedron 33: 1809
Gutman I, Cyvin SJ (1987) Chem. Phys. Letters 136: 137
Gutman I, Cyvin SJ (1988a) J. Serb. Chem. Soc. 53: 391
Gutman I, Cyvin SJ (1988b) Match 23: 175
Gutman I, El-Basil S (1984) Z. Naturforsch. 39a: 276
Gutman I, El-Basil S (1985) J. Serb. Chem. Soc. 50: 25
Gutman I, Herndon WC (1975) Chem. Phys. Letters 34: 387
Gutman I, Hosoya H, Yamaguchi T, Motoyama A, Kuboi N (1977) Bull. Soc. Chim. Beograd 42: 503
Gutman I, Mallion RB, Essam JW (1983) Molec. Phys. 50: 859
Gutman I, Marković S, Marinković M (1987) Match 22: 277
Gutman I, Obenland S, Schmidt W (1985) Match 17: 75
Gutman I, Petrović S (1983) Chem. Phys. Letters 97: 292
Gutman I, Polansky OE (1986) *Mathematical Concepts in Organic Chemistry*. Springer, Berlin Heidelberg New York
Gutman I, Randić M (1979) Chem. Phys. 41: 265
Hahn JH, Zenobi R, Bada JL, Zare RN (1988) Science 239: 1523
Hall GG (1973) Int. J. Math. Educ. Sci. Technol. 4: 233
Hall GG (1988) Theor. Chim. Acta 73: 425
Ham NS (1958) J. Chem. Phys. 29: 1229
Hara O, Tanaka K, Yamamoto K, Nakazawa T, Murata I (1977) Tetrahedron Letters: 2435
Hara O, Yamamoto K, Murata I (1977) Tetrahedron Letters: 2431
Harary F (1967) in: Harary F (ed) *Graph Theory and Theoretical Physics*. Academic Press, London, p 1
Harary F, Harborth H (1976) J. Comb. Inf. & System Sci. 1: 1
Harary F, Read RC (1970) Proc. Edinburgh Math. Soc. (Ser. II) 17: 1
Harary F, Schwenk AJ (1973) Discrete Math. 6: 359
Harvey RG, Geacintov NE (1988) Acc. Chem. Res. 21: 66
He WC, He WJ (1987) in: King RB, Rouvray DH (eds) *Graph Theory and Topology in Chemistry*. Elsevier, Amsterdam, p 484
He WC, He WJ, Cyvin BN, Cyvin SJ, Brunvoll J (1988) Match 23: 201
He WJ, He WC (1986) Theor. Chim. Acta 70: 43
He WJ, He WC, Wang QX, Brunvoll J, Cyvin SJ (1988) Z. Naturforsch. 43a: 693
Heilbronner E (1962) Helv. Chim. Acta 45: 1722
Hendel W, Khan ZH, Schmidt W (1986) Tetrahedron 42: 1127
Herbstein FH, Schmidt GMJ (1954) J. Chem. Soc.: 3302
Herndon WC (1973) J. Am. Chem. Soc. 95: 2404
Herndon WC (1974a) J. Am. Chem. Soc. 96: 7605
Herndon WC (1974b) Thermochim. Acta 8: 225
Herndon WC (1975) J. Org. Chem. 40: 3583
Herndon WC (1976) J. Am. Chem. Soc. 98: 887
Herndon WC (1980) Israel J. Chem. 20: 270
Herndon WC, Ellzey ML (1974) J. Am. Chem. Soc. 96: 6631
Herndon WC, Hosoya H (1984) Tetrahedron 40: 3987
Herndon WC, Párkányi C (1976) J. Chem. Educ. 53: 689
Hess BA, Schaad LJ, Herndon WC, Biermann D, Schmidt W (1981) Tetrahedron 17: 2983
Hosoya H (1971) Bull. Chem. Soc. Japan 44: 2332

Hosoya H (1986) Comp. & Math. with. Appl. 12B: 271
Hosoya H, Yamaguchi T (1975) Tetrahedron Letters: 4659
Jenny W, Peter R (1965) Angew. Chem. 77: 44
Joela H (1975) Theor. Chim. Acta 39: 241
Klarner DA (1965) Fibonacci Quart. 3: 9
Klein DJ, Schmalz TG (1989) Int. J. Quantum Chem. 35: 373
Klein DJ, Trinajstić N (1989) Pure Appl. Chem. in press
Knop JV, Müller WR, Szymanski K, Trinajstić N (1986a) J. Comput. Chem. 7: 547
Knop JV, Müller WR, Szymanski K,‚Trinajstić N (1986b) Match 20: 197
Knop JV, Szymanski K, Jeričević Ž, Trinajstić N (1983) J. Comput. Chem. 4: 23
Knop JV, Szymanski K, Jeričević Ž, Trinajstić N (1984) Match 16: 119
Lang KF, Eigen I (1967) Fortsch. Chem. Forsch. 8: 91
Léger A, d'Hendecourt L, Boccara N (eds) (1987) *Polycyclic Aromatic Hydrocarbons and Astrophysics.* Reidel, Dordrecht
Lempka HJ, Obenland S, Schmidt W (1985) Chem. Phys. 96: 349
Longuet-Higgins HC (1950) J. Chem. Phys. 18: 265
Lovász L, Plummer MD (1986) *Matching Theory.* North-Holland, Amsterdam
Lunnon WF (1972) in: Read RC (ed) *Graph Theory and Computing.* Academic Press, New York, p 87
Mikeš F, Boshart G, Gil-Av E (1976) J. Chem. Soc. Chem. Commun. 99
Müller E, Müller-Rodloff I (1935) Liebigs Ann. Chem. 517: 134
Murdoch J, Geissman TA (1967) Am. Miner. 52: 611
Obenland S, Schmidt W (1987) in: Léger A, d'Hendecourt L, Boccara N (eds) *Polycyclic Aromatic Hydrocarbons and Astrophysics.* Reidel, Dordrecht, p 165
Ohkami N, Hosoya H (1983) Theor. Chim. Acta 64: 153
Ohkami N, Motoyama A, Yamaguchi T, Hosoya H, Gutman I (1981) Tetrahedron 37: 1113
Pauling L (1960) *The Nature of the Chemical Bond.* Cornell Univ. Press, Ithaca, pp 234—239
Pauling L (1980) Acta Cryst. B36: 1898
Pauling L (1987) Nature 325: 396
Pauling L, Brockway LO, Beach JY (1935) J. Am. Chem. Soc. 57: 2705
Pauncz R, Cohen A (1960) J. Chem. Soc.: 3288
Pering KL, Ponnamperuma C (1971) Science 173: 237
Polansky OE, Derflinger G (1967) Int. J. Quantum Chem. 1: 379
Polansky OE, Rouvray DH (1967a) Match 2: 63
Polansky OE, Rouvray DH (1976b) Match 2: 91
Polansky OE, Rouvray DH (1977) Match 3: 97
Randić M (1974) Tetrahedron 30: 2067
Randić M (1975a) Croat. Chem. Acta 47: 71
Randić M (1975b) Tetrahedron 31: 1477
Randić M (1976) Chem. Phys. Letters 38: 68
Randić M (1977a) J. Am. Chem. Soc. 99: 444
Randić M (1977b) Tetrahedron 33: 1905
Randić M (1980) Int. J. Quantum Chem. 17: 549
Randić M (1982) J. Phys. Chem. 86: 3970
Randić M, Nikolić S, Trinajstić N (1987) in: King RB, Rouvray DH (eds) *Graph Theory and Topology in Chemistry.* Elsevier, Amsterdam, p 429
Randić M, Solomon V, Grossman SC, Klein DJ, Trinajstić N (1987) Int. J. Quantum Chem. 32: 35
Sachs H (1984) Combinatorica 4: 89
Schaad LJ, Hess BA (1972) J. Am. Chem. Soc. **94**: 3068
Schaad LJ, Hess BA (1982) Pure Appl. Chem. 54: 1097
Schmidt W (1987) in: Léger A, d'Hendecourt L, Boccara N (eds) *Polycyclic Aromatic Hydrocarbons and Astrophysics.* Reidel, Dordrecht, p 149
Searle CE (1986) Chem. Britain 22: 211
Searle CE, Waterhouse JAH (1988) Chem. Britain 24: 771
Simpson WT (1953) J. Am. Chem. Soc. 75: 597

Simpson WT (1956) J. Am. Chem. Soc. 78: 3585
Smith FT (1961) J. Chem. Phys. 34: 793
Somayajulu GR, Zwolinski BJ (1974) J. Chem. Soc. Faraday II 70: 1928
Staab HA, Diederich F (1983) Chem. Ber. 116: 3487
Staab HA, Sauer M (1984) Liebigs Ann. Chem.: 742
Stein SE (1978) J. Phys. Chem. 82: 566
Stein SE, Golden DM (1977) J. Org. Chem. 42: 839
Stein SE, Golden DM, Benson SW (1977) J. Phys. Chem. 81: 314
Stojmenović I, Tošić R, Doroslovački R (1986) in: Tošić R, Acketa D, Petrović V (eds)
    Proceedings of the Sixth Yugoslav Seminar on Graph Theory Dubrovnik 1985. Univ. Novi
    Sad, Novi Sad, p 189
Swinborne-Sheldrake R, Herndon WC, Gutman I (1975) Tetrahedron Letters: 755
Tošić R, Kovačević M (1988) J. Chem. Inf. Comput. Sci. 28: 29
Trinajstić N (1983) *Chemical Graph Theory*. CRC Press, Boca Raton
Ullmanns Encyklopädie der technischen Chemie, Vol 22 (1982) Verlag Chemie, Weinheim
Vogler H, Trinajstić N (1988) J. Mol. Struct. (Theochem) 164: 325
Wheland GW (1933) J. Chem. Phys. 3: 356
Wheland GW (1944) *The Theory of Resonance and Its Application to Organic Chemistry*.
    Wiley, New York, p 203
Wheland GW (1955) *Resonance in Organic Chemistry*. Wiley, New York, p 394
Zhang F, Chen R (1986) Match 19: 179
Živković T (1986) Int. J. Quantum Chem. 30: 591

Subject Index

H.-G. Franck, J. W. Stadelhofer, Frankfurt, FRG

Industrial Aromatic Chemistry

Raw Materials – Processes – Products

1988. XIV, 486 pp. 206 figs. 88 tabs. 720 structural formulas. ISBN 3-540-18940-8

From the contents: The Nature of the Aromatic Character. – Base Materials for Aromatic Chemicals. – Production of Benzene, Toluene and Xylenes. – Production and Uses of Benzene Derivatives. – Production and Uses of Toluene Derivatives. – Production and Uses of Xylene Derivatives. – Polyalkylated Benzenes. – Naphthalene – Alkylnaphthalenes and Other Bicyclic Aromatics. – Anthracene. – Further Polynuclear Aromatics. – Production and Uses of Carbon Products from Mixtures of Condensed Aromatics. – Aromatic Heterocycles – Toxicology/ Environmental Aspects. – The Future of Aromatic Chemistry.

The book is an adapted and updated translation of the German edition.

Springer-Verlag
Berlin Heidelberg
New York London
Paris Tokyo
Hong Kong

Deutsche Ausgabe:
H.-G. Franck, J. W. Stadelhofer:
Industrielle Aromatenchemie.
1987. ISBN 3-540-18146-6

Springer

I. Gutman, Kragujevac, Yugoslavia;
O. E. Polansky, Mülheim a. d. Ruhr,
FRG

Mathematical Concepts in Organic Chemistry

1986. X, 212 pp. 28 figs.
ISBN 3-540-16235-6

Contents: Introduction. – Chemistry
and Topology: Topological Aspects in
Chemistry. Molecular Topology. –
Chemistry and Graph Theory: Chemi-
cal Graphs. Fundamentals of Graph
Theory. Graph Theory and Molecular
Orbitals. Special Molecular Graphs. –
Chemistry and Group Theory: Funda-
mentals of Group Theory. Symmetry
Groups. Automorphism Groups. Some
Interrelations between Symmetry and
Automorphism Groups. – Special
Topics: Topological Indices. Thermo-
dynamic Stability of Conjugated Mole-
cules. Topological Effect on Molecular
Orbitals. – Appendices 1–6. – Litera-
ture. – Subject Index.

Springer-Verlag
Berlin Heidelberg
New York London
Paris Tokyo
Hong Kong

Springer